THE ROUGH GUIDE to the

iPhone

Peter Buckley

ROUGH
GUIDES

Credits

The Rough Guide to the iPhone

Text, design and layout:
Peter Buckley & Duncan Clark
Proofreading: Susanne Hillen
Production: Rebecca Short

Rough Guides Reference

Editors: Kate Berens,
Tom Cabot, Tracy Hopkins,
Matt Milton, Joe Staines
Director: Andrew Lockett

Apple hardware images courtesy of Apple UK

Acknowledgements

The author would like to thank everyone at Rough Guides, above all Sean Mahoney, who gallantly stood in line for 14 hours on the day of the iPhone's launch to secure the hardware used to write the first edition of this book. Also Andrew Lockett and John Duhigg for signing up the project in record time.

Publishing information

This fourth edition published February 2012 by
Rough Guides Ltd, 80 Strand, London WC2R 0RL

Penguin Group (USA), 375 Hudson Street, NY 10014, USA
Penguin Group (India), 11 Community Centre, Panchsheel Park, New Delhi 110017, India
Penguin Group (Australia), 250 Camberwell Road, Camberwell, Victoria 3124, Australia
Penguin Group (New Zealand), 67 Apollo Drive, Rosedale, Auckland 0632, New Zealand

Rough Guides is represented in Canada by Tourmaline Editions Inc.,
662 King Street West, Suite 304, Toronto, Ontario M5V 1M7

Printed and bound in Singapore by Toppan Security Printing Pte. Ltd.

Contents

Music & video

Reading

The Internet

Navigation

App essentials

Extras

Introduction

The iPhone was probably the most keenly anticipated consumer product of all time. After years of rumours, predictions and discussions, Apple finally previewed the device in early 2007. That was six months before the iPhone went on sale in the US and nearly a year before it would be available in Europe, but photos of the gadget were nonetheless all over newspaper front pages around the world the following morning. It was the sort of media coverage that no other company producing its first phone could have dreamed of. But then no other company had created the iPod or the Mac.

In the months after the preview, traders sold iPhone-related web addresses on eBay for huge sums, accessory manufacturers scrambled to produce add-ons for a device they'd never actually seen, and bloggers dissected Steve Jobs' presentation in minute detail for clues about what the iPhone would or wouldn't be able to do.

By the time the phone actually went on sale, things had reached fever pitch. Apple and AT&T stores across the US saw lines of early adopters camping on the streets to ensure that they wouldn't miss out on release day and have to wait a couple of weeks for restocks. Savvy students hired themselves out as line-standers, and websites pointed iPhone campers in the direction of the most conveniently located toilets, restaurants and DVD rental stores.

That was the Friday. By the end of the weekend, more than half a million phones had been sold, with another half-million or so flying off the shelves the following week.

But was all the excitement justified? Not according to some commentators, who described such scenes as examples of mass hysteria – triumphs of PR spin. Phones that do email, music and web browsing had existed for years, they pointed out, and many of them offered greater functionality or a broader feature array than the iPhone. And at a lower price, too.

But that was to miss the point. There had indeed been many smartphones before the iPhone, but they were consistently ugly, fiddly, counterintuitive and not ergonomic. They looked and felt as if they were designed by people interested in dry functionality rather than whether a device is pleasurable to use – people who create "solutions" rather than everyday products. Such phones were aimed primarily at businesses and had names reminiscent of the corporate computer systems they were designed to work with – the Motorola MC70, for instance, or Palm Treo 700WX.

By comparison, the iPhone was, from day one, nicely designed, simple to use, and – for the most part – did what you actually wanted it to do. Its key innovations have related not to "features" as such but to its human interface – a screen that lets you easily flick through long lists of contacts or zoom in on a webpage with the touch of a finger or two; a sensor that lets you rotate the device to make a landscape picture fill the screen or control a video game.

Once people had seen these features in action, the hype was generated from the bottom up. All Apple did was to put out a few ads – there was no guerrilla marketing campaign or PR offensive. (Indeed, as all technology journalists know, it's difficult to even get through to Apple PR, let alone get someone to return your call or provide products for review.)

The obsession never went away: in October 2011 the launch of the fifth-generation iPhone – the 4S – saw punters queuing outside stores once again. The iPhone 4S sold a remarkable four millions units in the first weekend, making it the fastest-selling phone of all time.

None of which is to say that the iPhone is beyond criticism. It has its faults and limitations, of course. And you could make a reasonable case that so much buzz over a mobile phone, however attractive it may be,

is an example of modern society's unhealthy obsession with expensive brand-name goods and techie gadgets.

That may well be, but mobile phones are here to stay. And if we're going to have them, we may as well have ones that are good to use and long-lasting – and which include music and camera functions so that we don't end up buying and carrying around multiple devices. The iPhone checks all of these boxes, and plenty more besides.

About this book

Whether you're thinking about buying an iPhone or already have one, this book is for you. It covers the whole topic, from questions you might want answering before you buy, and advice on importing numbers from an old handset, through to advanced tips and tricks and pointers to some of the best apps.

This book was written using an iPhone 4S running iOS 5.0. (To see which version of the software you're running, click Settings > General > About.) If you're running a later version of the OS then you may well come across features not covered here, though the majority of what's written will still apply. If you have an earlier version of the software or an older iPhone, you may be missing a few of the features described here but, again, the majority of the book should still be relevant.

Primer

01
FAQs

Everything you ever wanted to know but were afraid to ask

The big picture

What's an iPhone?

An iPhone is a smartphone – in other words, a mobile phone that doubles as a handheld computer, complete with web browsing, email, music playback and the ability to run applications, or "apps". The device is produced by Apple, manufacturers of the iPod, iPad and Mac computers.

After years of speculation, and to great media fanfare, the iPhone was launched in the US and UK in 2007, with an updated version released each summer since.

How does it compare to other smartphones?

There are lots of smartphones on the market, many of which offer a similar feature list to that of the iPhone, including web browsing, email and downloadable apps. These iPhone competitors are made by lots of different companies, but most of them use the same underlying software – a system called Android, produced by online giant Google. Android phones vary a lot in terms of price and quality – and the degree

to which the phone manufacturer customizes the software. Whether you choose an iPhone or Android phone is really down to personal preference. Interestingly, the rivalry between the two platforms and their users is beginning to mimic the rivalry between Mac and Windows PC users in the home computing space. For the full story, check out *The Rough Guide to Android Phones and Tablets*.

What's an app?

App is short for application – a piece of software designed to fulfil a particular function. If you're used to a PC, an app is basically the same as a program. On the iPhone an app might be anything from a game, a word processor or a retro-styled alarm clock through to a version of a popular website with extra features added specially for the iPhone.

There are hundreds of thousands of apps available to download from the App Store (see p.68), some of which have been downloaded and installed on millions of iPhones.

How does the iPhone compare to the iPod?

The iPhone *is* an iPod, in addition to being a phone and an Internet device. It can do everything an iPod can do: play music, videos and podcasts, display album artwork using "Cover Flow" (see p.173), and create playlists based on a single track using the "Genius" feature (see p.168). The main advantage of a traditional iPod is storage space. At the time of writing, iPhones offer up to 64 gigabytes of space, while the largest iPod model has 160 gigabytes.

How does it compare to a computer for web browsing and email?

No pocket-sized Internet device can match a computer with a large screen, mouse and full-sized keyboard, but the iPhone comes about as close as you could hope. Web browsing, in particular, is very well handled, with a great interface for zooming in and out on sections of a webpage. However, when you're out and about, you'll find the access slow compared to a home broadband connection. (More on this later.)

Can it open and edit Word and Excel docs?

Out of the box, the iPhone can open, read and forward Word, Excel and PowerPoint docs sent by email or found online, but it doesn't allow you to edit them. To do this, you'll need to download a suitable app.

What's iOS and OS X?

All computers have a so-called operating system – the underlying software that acts as a bridge between the hardware, the user and the apps. The standard operating system on PCs is Windows; on Macs, it's OS X.

The operating system on the iPhone – known as iOS – is based on OS X, though you wouldn't know it, because it's slimmed down and specially designed to make the most of the phone's small touch-screen interface.

Is the onscreen keyboard easy to use?

Apple are very proud of the iPhone's touch-screen keyboard and the accompanying error-correcting software that aims to minimize typos. In general, reviewers and owners alike have been pleasantly surprised at how quickly they've got used to it. Inevitably, however, it's not to everyone's taste. When using many applications you can rotate the iPhone through ninety degrees to use a bigger version of the keyboard in landscape mode.

Do I need a computer to use an iPhone?

No. Originally iPhones could only be activated with a Mac or PC but that isn't true any more thanks to iCloud (see opposite). However, a computer is still essential for some iPhone tasks. For example, if you want to copy a CD onto your iPhone, you'll first need to copy it onto a computer with a CD drive. You can, however, download music and video files straight to the iPhone from the iTunes Store (see p.161).

How does the phone connect to the computer?

The two can communicate via your home wireless network or using a USB cable.

What's iTunes?

iTunes is a piece of software produced by Apple for Macs and PCs. Its main function is as a tool for importing, downloading and managing audio and video – including creating playlists, importing CDs, subscribing to podcasts and buying music, film and TV shows from Apple. However, iTunes also functions as the hub for selectively moving content from your computer – such as music, video, podcasts and photos from your archive – across to your iPhone. iTunes can also be used for syncing contacts, calendars and other information with your computer – though these tasks can also be taken care of by iCloud.

What's iCloud?

iCloud is Apple's service for synchronizing content and settings between multiple devices – including iPhones, iPads and computers. Launched in late 2011, iCloud allows you to ensure, for example, that when you add a phone number to your iPhone it will also appear in the Address Book on your Mac, or when you add new music to your iPad that it will also be available on your iPhone. iCloud is free to use, though you have to pay if you require more than five gigabytes of online storage space (not including items purchased from Apple, which are stored for free). See p.40 for more information.

This is the cloud the way it should be: automatic and effortless. iCloud is seamlessly integrated into your apps, so you can access your content on all your devices. And it's free with iOS 5.
Learn more ›

What's MobileMe?

MobileMe was an Apple subscription service that provided a suite of online tools in return for an annual fee – including various tools for iPhone users. However, the service was discontinued for new users in 2011, with the launch of iCloud. Existing MobileMe users will have access to some services into 2012, before being fully transitioned to iCloud.

Does the iPhone contain a hard drive?

No. Like the iPod Touch and some recent Mac laptops, the iPhone stores its information on flash memory: tiny chips of the kind found in digital camera memory cards and thumb drives. These have a smaller capacity than hard drives, but on the other hand they're less bulky, less power hungry and less likely to break if the device is dropped.

Does the screen scratch easily?

Though not invulnerable, recent iPhone models have screens that are surprisingly scratch resistant. Nonetheless, you might want to consider investing in an inexpensive screen-protecting film or some kind of wrap-around case (see p.270) to protect the screen when your phone is in a pocket or bag.

Does the iPhone present any security risks?

Not if you're sensible. There's a theoretical risk with any smartphone or computer that someone could "hack" it remotely and access any stored information. But the risk is extremely small – especially in the case of the iPhone which, by default, won't allow Bluetooth access from laptops or other phones. The only real risk – as with any phone – is that someone could steal your iPhone and access your private data or make expensive long-distance calls on your account. If you're worried about that possibility, the best defence is to password-protect your iPhone – see p.50 to find out how.

What's the iPhone's battery life like?

The iPhone has relatively good battery life compared to its competitors, but – as with all mobile phones and computers – the amount of time that the battery lasts depends entirely on what you're doing with the iPhone at any one time (see p.262 for full details).

Despite its respectable battery performance, the iPhone – like the iPod in previous years – has sometimes been criticized in the press and elsewhere for the fact that the battery gradually dies over time and can only be replaced by Apple for a substantial fee ($79/£55 at the time of writing). To be fair, however, *all* rechargeable batteries deteriorate over time and eventually die. The only difference with iPhones is that the replacements cost more than some other phones – and that you can't fit them yourself (at least not in theory). The high cost can partly be explained by the nature of the battery. Few pocket devices offer hours of video playback or such large, bright screens. That type of performance is comparable to a laptop – and laptop batteries are even more expensive. As for sending the device back to Apple, this is irritating, for sure, but it also means that the iPhone can be properly sealed and that it's free from the flimsy battery flaps that often get broken on other phones.

Although Apple don't recommend it, it's possible to source bargain iPhone batteries on the Internet and fit them yourself, or pay someone else to do it. As for the longevity of the battery, expect to replace it after around two to four years, depending on how much you drain it each day. For more information about the battery, see p.262.

Is it possible to type with one hand?

Yes, this is perfectly possible: your fingers hold the body of the device and your thumb can access the screen. However, typing is faster with two hands.

Will the touch screen work with gloves on?

No – unless you wear fingerless gloves or buy some special ones designed for use with an iPhone. There are many brands available, one example being dotsgloves.com. You could also consider a stylus; the pick of the bunch is the Pogo (tenonedesign.com/stylus). Or you could take your lead from Korean iPhone users, who swear by a particular brand of snack sausage (tinyurl.com/yldwb38).

Phone issues

If I get an iPhone, do I have to change network and get a new contract?

It depends. There are two ways to buy an iPhone: either "unlocked" directly from Apple, or as part of a new or repeat contract from a network that offers special iPhone plans. The unlocked option means spending two or three times as much money upfront, but the phone won't be locked to any particular network or contract. Therefore you can use it with any GSM-based phone carrier on a contract of your choice – including prepaid/pay-as-you-go. With an unlocked phone you can also swap more than one SIM card around – useful if you want to be able to use your phone with a local carrier when travelling overseas.

However, for most people, the best option is to get an iPhone as part of a new contract – either with your existing carrier as part of an upgrade or with a different one. That way, the upfront costs is significantly reduced and the price of the phone is subsidized by the carrier. The only problem with this option is that many users are locked into an existing phone contract that may not expire for some months or even years. In that situation, the two options are buying an unlocked iPhone or starting a new contract and paying whatever it takes to get out of the old one.

Either way, you'll usually get to keep your current phone number.

A MicroSIM is just a normal SIM card cut down to a smaller size – as can be seen in this O2 SIM which can be used in traditional or Micro format

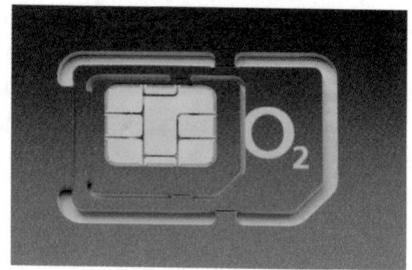

Can I use my current SIM card?

Generally not. For one thing, your existing card may not fit the iPhone, which has used a smaller format called MicroSIM since the launch of model 4 – though some phone shops will cut down full-sized SIMs to MicroSIM size. The second barrier is that if the iPhone is locked to a particular network or contract, it will reject SIMs from other networks or contracts.

Naturally, it didn't take geeks long to figure out how to "unlock" locked iPhones for use with any network and suitably sized SIM card. Sites such as iphoneunlocking.org.uk offer downloadable software especially for the purpose. Unlocking an iPhone allows certain benefits – such as using a foreign MicroSIM when abroad, or switching to an unsupported carrier offering better call plans. However, there are also downsides – most obviously, the fact that you'll invalidate your warranty. Certain unlocking solutions may also disable some functions and make it impossible (or risky) to install future software updates from Apple.

What about "jailbreaking"?

Jailbreaking is the process of adapting the iPhone's operating system to allow it to run third-party software that's not been approved by Apple. Due to the existence of the App Store (see p.69), and the wealth of applications available, there isn't much reason to jailbreak an iPhone any more, but if you're really keen, there's plenty of information available online.

Can I use the iPhone overseas?

Yes, the iPhone will work for voice calls almost anywhere in the world. However, before making calls overseas, some US-based iPhone owners – and UK customers on certain networks – may first need to activate international roaming with their carrier. Either way, be aware that making and receiving calls overseas can be very expensive – be sure to check your carrier's international tariffs to avoid getting stung.

Using data when you're abroad is even more expensive than making calls abroad – so much so that for Internet-based features and apps such as web, email, maps and so on, it's usually best to stick to Wi-Fi hotspots where you can get online for free (or at the cost of accessing the hotspot). In many countries it's possible to connect via the local phone network as well, but be prepared for some savage fees. When you're paying up to $19 per megabyte, then a week or so of using email, web or maps can quickly add up to hundreds of dollars of charges.

If you can stomach such high costs, you can switch on Data Roaming within Settings > General > Network. But if you do so, be careful. A single YouTube video may be dozens of megabytes – and could, therefore, cost as much as a dozen cinema tickets. Also bear in mind that, with Data Roaming turned on, your phone will most likely receive some data automatically, even if you don't consciously use any Internet-based apps. To minimize the risk of inadvertent downloading, switch off automatic fetching of emails or receiving of Push notifications within Settings > Mail, Contacts, Calendar > Fetch New Data.

To avoid roaming charges, you could consider trying to get your iPhone unlocked (see p.18), which would allow you to use it with a local pay-as-you-go SIM or MicroSIM from the country you're visiting. But unlocking presents various problems, as already discussed, and when a foreign SIM is in place you won't be able to receive calls to your usual number. All told, if you travel to one country a lot, it might be simpler to buy an inexpensive prepaid phone in that country and arrange to have calls to your regular iPhone number forwarded. See p.94 to find out how.

Will I be able to get the contact numbers off my old phone and onto the iPhone?

In general, yes, but the process depends on your old phone and your computer. See p.101 for more information.

Internet issues

Does the iPhone offer fast Internet access?

That depends where you are. If you're within range of a Wi-Fi network, as found in many homes, offices and cafés (and across some entire city centres), then your Internet access will typically be pretty speedy; not quite as fast in practice as with a Mac or PC, but not far off.

If, on the other hand, you're out and about – walking down the street, say, or sitting on a bus – then the iPhone will usually have to access the Internet via a mobile phone network. Where you can get access to a 3G signal, the speed of your connection will be passable, though in most areas it's still nowhere near as fast as a home broadband connection. Whenever 3G isn't available, your iPhone will try to connect to the slower EDGE and GPRS networks (see below).

Believe it or not, another factor worth considering when you're out and about is your own speed of travel. On a train or in a moving car, you might find that your connection speed is slower than when stationary. This is because the device has to accommodate a constantly shifting relationship to nearby signal masts, making it hard for the iPhone to maintain a coherent stream of data to and from the Internet.

EDGE, 3G... what's all that about?

Over time, the technology used to transmit and receive calls and data from mobile phones has improved, allowing greater range and speed. Of all the network technologies widely available at present, 3G (third generation) is the most advanced, allowing Internet access at speeds comparable to home broadband connections – in theory, at least.

All iPhones, except the first-generation model, can make use of 3G. Where that's not available, the iPhone falls back on the next best option, which is a network known as EDGE (Enhanced Data rates for GSM Evolution, to give it its rather grand full name). EDGE is much slower than 3G, but it does have one advantage: accessing EDGE uses less power than accessing 3G. For this reason, if you're low on power, haven't got a charger and desperately need to get online, you may want to turn off 3G for a while, which is easily done within Settings > General > Network.

Will all websites work on an iPhone?

The iPhone features a fully fledged web browser – Safari – which will work with the overwhelming majority of websites. The main catch is that certain special types of web content won't display:

• **Flash animations** Widely used for banner ads (no great loss there), interactive graphics and a few entire websites.

• **Java applications** Not to be confused with Javascript (which works fine on an iPhone), Java is used for certain online programs such as calendar tools or broadband speed tests.

• **Some music and video** Music and video clips will usually work fine, assuming you have a reasonable connection – though quite a few are encoded using Flash. These ones won't show up.

What's Safari?

Safari is the web browser built into the iPhone. It's a streamlined version of the browser shipped on all Macs. Just before the release of the iPhone, in mid-2007, Apple released a version of Safari for PC users.

Can the iPhone sync my bookmarks?

Yes, it works automatically if you use Safari or Internet Explorer, and it's possible to sync with other browers using a service such as Xmarks. See p.213 for more information.

Will the iPhone work with my email account?

For a personal email account, almost certainly. The iPhone works with AOL, Yahoo!, iCloud, MobileMe and Gmail with just a username and password. If you have a different email provider, it's still easy to set up – and if you use Outlook on a PC or Mail on a Mac, the iPhone will even sync your account details so you don't have to worry about entering server details and the like.

The only time you may encounter a problem is when setting up a work email account, because some companies block the ability to access email on non-work devices. The only way to be sure is to ask. For more on setting up and using email, see p.117.

Can I make phone calls while on the Internet?

Yes, but only when your iPhone is connected to the Internet via a Wi-Fi network. You can't make calls and get online via EDGE or 3G simultaneously.

Can I use the iPhone to get my laptop or iPad online when out and about?

Yes, recent versions of the iPhone software have made it easy to share the phone's Internet connection – see p.66 for more details. Note, though, that on some phone networks you may have to pay an additional "tethering" fee to your mobile phone provider in order to get this feature to work.

Music, video and camera issues

How good is the iPhone camera? Could it replace my normal camera?

The iPhone's camera has got better with each model. In the earlier models, the camera was handy enough but not exactly high quality

– and less good than those found on other smartphones. With the iPhone 4S, however, the camera has taken a big leap forward, thanks to a higher resolution sensor and – crucially – better lenses. Of course, it can't compete with a digital SLR, but it does a decent job of substituting for most pocket-sized point-and-shoot digital cameras. Best of all, the iPhone 4S is capable of shooting high defintion (HD) video. See p.133 for more on the iPhone's camera.

Will the iPhone work with my iPod accessories?

Maybe. The Dock socket on the bottom of the iPhone is the same as the one on an iPod, so iPod accessories that connect via this socket should be able to connect to an iPhone. That doesn't necessarily mean they'll work in practice, however – especially if they are more than a couple of years old.

What's the sound quality like for music?

Pretty much the same as with an iPod. Tracks downloaded from the iTunes Store or imported from disc at the default settings sound marginally worse than CD quality. However, you're unlikely to notice any difference unless you do a side-by-side comparison through high-quality speakers or headphones.

Anyhow, this sound quality isn't fixed. When you import tracks from CD (or record them from vinyl) you can choose from a wide range of options, up to and including full CD quality. The only problem is that better-quality recordings take up more disk space, which means fewer tracks on your phone. The trade-off between quality and quantity is entirely for you to decide. For more information, see p.149.

Can I use my existing earphones or headphones?

Yes. However, if you use regular headphones or earphones you won't be able to use the mic and button that form part of the supplied earbuds. These allow you to answer calls and control music playback without getting the device out of your pocket. For details of some headphones that do work with the iPhone, turn to p.272.

I've never had an iPod. Is it a hassle to transfer my music from CD?

It certainly takes a while to transfer a large CD collection onto your Mac or PC, but not as long as it would take to play the CDs. Depending on your computer, it can take just a few minutes to transfer the contents of a CD onto your computer's hard drive – and you can listen to the music, or work in other applications, while this is happening. Still, if you have more money than time, there are services that will take away your CDs and rip them into a well-organized collection for around $1/£1 per CD.

PodServe podserve.co.uk (UK; currently London area only)
DMP3 dmp3music.com (US)
RipDigital ripdigital.com (US)

Once your music is on your PC or Mac, it only takes a matter of minutes to transfer even a large collection across to the iPhone; and subsequent transfers are even quicker, since only new or changed files are copied over.

Can the iPhone download music directly?

Yes, this is possible using the iTunes Store, see p.161.

Is downloading legal?

Yes, as long as you use a legal store such as iTunes. As for importing CDs and DVDs, the law is, surprisingly, still a tiny bit grey in many countries, but in practice no one objects to people importing their own discs for their own use. What could put you on the wrong side of the law is downloading copyrighted material that you haven't acquired legitimately – and, of course, distributing copyrighted material to other people.

What is DRM?

DRM (digital rights management) is the practice of embedding special code in audio, video or eBook files to limit what the user can do with those files. For example, ePub titles downloaded from the iBookstore (see p.193) will only be readable using the iBooks app on an iPhone, iPad or the iPod Touch. Most music and video downloaded from the iTunes Store features FairPlay DRM, which allows the files to be played on up to five computers that have been authorized to work with the iTunes account used to download them. These files can, however, be used on as many iPhones, iPods and iPads as you want.

02

Buying an iPhone

Which model? Where from?

For most people, buying an iPhone will simply involve walking into a store and picking whichever capacity model they can afford. But, for more cautious buyers, a few questions may need to be answered first. How much storage space do you really need, for example? And would it be better to wait for a next-generation iPhone?

How much space do you need?

At the time of writing, the latest iPhone – the 4S – is available with either 16, 32 or 64 gigabytes of storage space. The amount of space you need depends on the number of songs, photos, movies, apps, podcasts and email attachments you want to be able to store at any one time.

True storage capacity

The first thing you should know is that your iPhone may offer slightly less space than you expect. All computer storage devices are in reality about 7 percent smaller than advertised. The reason is that hardware manufacturers use gigabyte to mean one billion bytes, whereas in computing reality it should be 2^{30}, which equals 1.0737 billion bytes. This is a bit of a scam, but everyone does it and no one wants to break the mould.

Moreover, a few hundred megabytes of the remaining 93 percent of space is used to store the iPhone's operating system, applications and firmware. All told, then, you can expect to lose a decent chunk of space before you load a single video, song or app:

Advertised capacity	Real capacity	Actual free space
64GB	59.6GB	56.8GB
32GB	29.8GB	28.4GB
16GB	14.9GB	14.5GB

Checking your current data needs

If you already use iTunes to store music and video, then you can easily get an idea of how much space your existing collection takes up. Click Music, Movies, Podcasts or any playlist on the left of the screen and the bottom of the iTunes window will reveal the total disk space each one occupies.

As for photographs, the size of the images on your computer and the amount of space they occupy there bears little relation to the space that the same images would take up on the iPhone. This is because when the photos are copied to your phone (see p.138) they are resized and optimized for use on the iPhone's screen. As a guide, 3000 images will take up around 1GB on the phone.

How big is a gig?

A gigabyte (GB) is, roughly speaking, the same as a thousand megabytes (MB) or a million kilobytes (KB). Here are some examples of what you can fit in each gigabyte.

		1GB =
Music	at 128 kbps (medium quality)	250 typical tracks
	at 256 kbps (high quality)	125 typical tracks
	at 992 kbps (CD quality)	35 typical tracks
Audiobooks	at 32 kbps	70 hours
Photos		3000 photos
Video		2.5 hours

To buy or to wait?

When shopping for any piece of computer equipment, there's always the tricky question of whether to buy the current model, or hang on for the next version, which may be better *and* less expensive. In the case of Apple products, the situation is worse than normal, because the company is famously secretive about plans to release new or upgraded hardware.

Unless you have a friend who works in Apple HQ, you're unlikely to hear anything from the horse's mouth until the day a new product appears. So, unless a new model came out recently, there's always the possibility that your new purchase will be out of date within a few weeks. About the best you can do is check out sites where rumours of new models are discussed. But don't believe everything you read …

Mac Rumors buyersguide.macrumors.com
Apple Insider appleinsider.com
Think Secret thinksecret.com

Where to buy?

Unlike iPods and Macs, in some countries the iPhone is only available direct from Apple or the partner phone carrier in your country. The price will usually be the same – or so close that it's unlikely to be a significant factor.

Buying from a high-street store means you'll get the phone immediately; ordering online you can expect a week's wait for delivery. Another advantage of visiting a store is that you get to see the thing in the flesh and try the various features before you buy. To find your nearest Apple Store, and to check for the availability of iPhones at various different branches, see:

Apple Stores apple.com/retail

Or to find the nearest branch of your carrier, check online for a store locator page. For example:

AT&T (US) www.att.com/storelocator/iphone
O2 (UK) o2.asymmetry.co.uk

New York City's Fifth Avenue Apple Store

What's in the box?

At the time of writing, the iPhone comes with a stereo headset with mic/button, a USB charging/sync cable, a charger and a polishing cloth. The first generation iPhone also included a Dock (see p.38) but, as with the iPod, this soon became an optional extra.

Used iPhones

Refurbished iPhones

Refurbished Apple products are either end-of-line models or up-to-date ones which have been returned for some reason. They come "as new" – checked, repackaged and with a full warranty – but are reduced in price by anything up to 40 percent. You'll find the Apple Refurb Store on the bottom-right of the Apple Store homepage. If there's something there you want, act quickly, as items are often in short supply.

Secondhand iPhones

Of course, it's possible to buy a used iPhone from an individual and get a new SIM card for it. However, many iPhones are locked to the network through which they were purchased, which means they'll only work with a SIM for the same network. If the phone has been unlocked, you'll be able to use any SIM card and network, but be aware that unlocking will probably have voided the warranty (see p.19).

As with all used electronic equipment, make sure that you see it in action before parting with any cash, but remember that this won't tell you everything. If an iPhone has been used a lot, for example, the battery might be on its last legs and soon need replacing, which will add to the cost (see p.262).

Recycling your old phone

You should never throw old mobile phones away. Not only do they contain chemicals that can be harmful to the environment when incinerated or sent to landfill; they also contain metals and other materials that can be recycled and used in new phones.

Anyhow, there are plenty of better options. Various groups will take old phones off your hands – even if they're broken. Apple provide addressed, postage-paid envelopes expressly for the purpose. To order one, visit:

Apple Recycling (US) apple.com/recycling
Apple Recycling (UK) apple.com/uk/recycling

However, Apple won't pay you for your old phone, whereas other services will. Old iPhones in particular can command relatively high prices, even if damaged. Try these sites first:

Mobile Valuer mobilevaluer.com (UK)
Cell for Cash cellforcash.com (US)

Alternatively, you could donate your old phone to charity. In the UK, for example, Oxfam can fund 83 school meals in the developing world for each handset dropped off at an Oxfam store or sent to: Oxfam Recycle Scheme, Freepost LON16281, London WC1N 3BR. In the US, try:

Phones4Charity phones4charity.org

Getting started

03

The basics

Setup, charging, syncing

It's simple to get started with a new iPhone, but it's worth spending a while getting to know your phone properly. This chapter covers all the basics – and also provides advice on synchronizing your phone with other devices.

Since the launch of the iPhone 4S in late 2011, setting up a new iPhone is as simple as turning it on and following a few simple instructions. However, if you don't already have an Apple ID, you'll need to set one up as part of the process. You can skip this step if you like but an Apple ID is required for downloading apps and music – and for using iCloud.

The only other thing you may need to know at the outset is how to import all your information and apps from your old iPhone. To do this, first sync your old iPhone with iTunes; then connect your new phone, choose Restore From Backup, choose your old iPhone's name and select the backup with today's date.

For help getting contacts off your old non-Apple phone, see p.100.

This computer has previously been synced with an iPhone or another iOS device.

◉ Set up as a new iPhone

○ Restore from the backup of: [ePhone ⬧]

Last Backed Up: Today 13:30

The iPhone at a glance

This diagram shows the iPhone 4S, which looks the same as the iPhone 4. Earlier models have a slightly different layout.

Headphone socket Takes standard stereo minijack plugs, but many require an adapter due to the narrow recess.

Front lens (video calls)

Silent ringer switch Toggles between Ring and Silent mode. You choose whether vibrate is on in one or both of these modes; see p.48.

Volume buttons Affects the ringtone (when nothing's happening) or speaker/ headphones (when you're on a call or playing music).

Notification area Swiping down from the top of the screen, as shown by the orange arrow, reveals a list of all your recent notifications, such as text messages and calendar alerts.

Speaker Comes on whenever you play music or video with no headphones plugged in. Also useful for calling – just press the onscreen Speaker button during a call. Anyone within a few feet can then get involved in the call.

Sleep/Wake Click once to sleep (will still receive calls); hold down for three seconds to power off (will no longer receive calls).

Rear lens (for photos and video)

MicroSIM tray Gives access to the MicroSIM card. To remove it, press the tiny circle with the tiny supplied tool – or use a paperclip

Status bar Displays the time and gives you feedback about your phone via various icons. These include the following:

📶 Phone signal level. Relates to calls rather than Internet. When abroad, you'll also see the name of the current carrier.

✈ Airplane mode on: phone, Wi-Fi and Bluetooth signals are disabled.

🔒 Phone locked.

▶ Music or podcast currently playing.

⏰ Alarm set. See p.237.

Data connection:
O GPRS (slowest)
E EDGE (faster)
3G 3G (fastest) depending on local reception...

📶 ... which is replaced by this when connected to Wi-Fi (see p.62). More bars means a stronger signal.

* Bluetooth is on. See p.65.

🔋 Battery charging.

🔋 Battery charged.

Home button Click to leave the app you are using and return to the last Home Screen you viewed. Whatever you're currently doing will be put on hold, so you can return to it later. Click again to go to the farthest left Home Screen. Click again to see the Search Screen (see p.52). Double click to see recently used apps (see p.71)..

Mic Works well enough from a few feet away when using speaker-phone.

Dock connector socket Takes the iPhone sync/charge cable – which is interchangeable with an iPod cable. The socket is also used for certain accessories.

Cables & docks

The iPhone comes with a charge/sync cable just like those used for the iPod. One end can attach directly to the iPhone or – if you choose to buy one – to a little stand known as a "Dock". The other end of the cable connects to a USB port – on a Mac or PC, on the supplied power adapter, or anywhere else.

The Dock

The Dock is a stand that makes it convenient to connect and disconnect the iPhone to computers, power sources, speakers or hi-fis. The Dock's connections can be left in place, so when you get home you simply drop your iPhone in and it's instantly hooked up, synced and charging. In addition, the Dock features a genuine line-out socket, as opposed to a headphone socket, so the sound quality is marginally improved when connecting to a hi-fi or speakers.

Official Apple iPhone docks are relatively expensive, though it's possible to buy cheaper third-party equivalents.

Using iPod Docks and cables

If you have a USB cable for an iPod, this should work fine with your iPhone – and vice versa. Some iPod Docks work too, though it depends on the model in question. If you have a Universal Dock, you can buy an inexpensive iPhone adapter for it. One difference between the iPhone Dock and an iPod Dock is that the former offers "special audio porting" – a rather grandiloquent name for some little holes in the base of the cradle that allow you to make use of the iPhone's speaker and microphone while the device is in the Dock.

Charging

To charge an iPhone, simply connect it to a USB port on a computer or power adapter. Note, though, that if you're using a computer, the USB port will need to be "powered". Most are, but some, such as those on keyboards and other peripherals, won't work. Also note that iPhones usually won't charge from a Mac or PC in sleep or standby mode.

When the iPhone is charging, the battery icon at the top-right of the screen will display a lightning slash. When it's fully charged this will change to a plug. When your phone is plugged in and not in use, you'll also see a large battery icon across its centre, which shows how charged it is at present. Like many mobile devices, iPhones use a combination of "fast" and "trickle" charging. This means it should take around two hours to achieve an 80 percent charge, and another two hours to get to 100 percent.

iPhone 4S battery life

Talk time: Up to 8 hours on 3G · Up to 14 hours on 2G	**Standby time:** Up to 200 hours
	Video playback: Up to 10 hours
Internet use: Up to 6 hours on 3G · Up to 9 hours on Wi-Fi	**Audio playback:** Up to 40 hours

If your iPhone's power becomes so low that the device can't function, you may well find that plugging it in will not revive it straight away. That's because the whole operation system needs to restart – just the same as when you restart a computer that's been properly switched off. Don't worry – it should come back to life after charging for a few minutes.

If you're in a hurry to charge, avoid syncing the iPhone with your computer while it's charging – this may slow the process. If a sync starts, you can cancel it in iTunes (see box overleaf) by clicking the ✖ icon next to the progress bar at the top of the iTunes window. For tips on maximizing your battery life see p.263.

Staying in sync

Although the iPhone works perfectly well on its own, there are lots of good reasons to sync it with other devices – such as your Mac or PC, or iPad. This way, all your photos, contacts and other information will be available on all your devices at any time – and safely backed up, too. There are two main ways to sync the iPhone with other devices, which you can mix and match as per your requirements. Neither is obligatory.

• **iCloud** Apple's online sync service, iCloud, provides seamless wireless syncing for the iPhone, iPad and iPod Touch – as well as Macs and PCs. iCloud keeps your contacts, calendars, recent photos, documents from many applications and other data synchronized across all your devices. It also backs up your iPhone and provides an online home for all the music, video and apps that you've bought from Apple, enabling you to easily install suitable apps on multiple devices. iCloud is free to use, though fees kick in if you need to sync and store more than a certain amount of data (excluding anything purchased from Apple, which is stored for free).

• **iTunes** Available for Mac and PC, iTunes is first and foremost an application for playing, arranging and downloading music, TV shows, films and podcasts. However, it has a secondary role as a control centre for your iPhone. It allows you to sync music, video, photos and other data between your computer and phone – and it also provides a fast way to rearrange apps on your Home Screens.

Syncing with iCloud

Launched with iOS 5 in autumn 2011, iCloud is an online service that supersedes previous Apple offerings such as MobileMe and .Mac. (The "cloud" of the title is a reference to cloud computing – a term used to describe technologies that make use of Web-based storage for data and applications.)

iCloud needs to be activated on each of the devices that you want to use it with. In each case you need to log in with an Apple ID.

> **→ TIP** In most cases it makes sense to use the same Apple ID for iCloud as you use for iTunes, though it's possible to have separate accounts – which can be useful if, for example, you share one Apple ID with your family for iTunes but each want to have separate contacts on your iPhone.

Where to find iCloud

• **On an iPhone or iPad** Log in when you first switch on or at any time later under Settings > iCloud. If you're using an older phone, you'll first need to upgrade to iOS 5 if you haven't already (see software update on p.259).

• **On a Mac** Open the iCloud section of System Preferences, which can be accessed via the Apple menu at the top-left of the screen. Note it will only work if you have a recent version of Mac operating system – OS X 10.7 ("Lion") or later. If you have Lion but can't find iCloud in System Preferences, run Software Update (also in the Apple menu) to make sure you have the latest version.

• **On a PC** If you haven't already done so, you'll need to download and install the iCloud Control Panel from apple.com/icloud/setup. Before downloading, check that your version of Windows is compatible and up to date.

• **On the web** You can access your iCloud account – and use various tools such as Find My iPhone (see below) – on the web at icloud.com.

Once logged in, your next task is to decide which parts of the iCloud service you want to use with each device. For many people, it makes sense to ignore the email option – this is only relevant if you're interested in setting up a new "@me.com" email address. But most of the other options are handy, including bookmarks, contacts, calendars, Photo Stream for recent pictures (see p.139) and Documents & Data, which allows iCloud-enabled apps to save files and other data directly to your iCloud account. On your iPhone, you'll probably also want to enable Find My iPhone, which allows you to find your phone on a map or make it beep – potentially very useful if you misplace it.

Once you've set up iCloud, everything should take care of itself: there's no need to manually tell it to sync.

Synchronizing with iTunes

If you've ever used an iPod, you'll already be familiar with iTunes – Apple's application for managing music, videos and podcasts, ripping CDs, and downloading music and video. Although in most cases iCloud is the best way to sync contacts, calendars and bookmarks, you'll still need to use iTunes to upload music that you didn't buy from Apple, any video content, and photos from your archive.

If you already use iTunes

Even if you already use iTunes, you may need to update to the latest version to get it to work with the iPhone. New versions come out regularly, and it's always worth having the latest. To make sure you have the most recent version, open iTunes and, on a Mac, choose Check for Updates… from the iTunes menu and, on a PC, look in the Help menu.

If you don't already use iTunes

All recent Macs have iTunes pre-installed. You'll find it in the Applications folder and on the Dock. Open it up and check for updates, as described above. If you have a PC, however, you'll need to download iTunes from Apple:

iTunes apple.com/itunes

Once you've downloaded the installer file, double-click it and follow the prompts. Either during the installation or the first time you run iTunes, you'll be presented with a couple of questions. Don't worry too much about these, but it's worth understanding what you're being asked…

• Yes, use iTunes for Internet audio content or
• No, do not alter my Internet settings
This is asking whether you'd like your computer to use iTunes (as opposed to whatever plug-ins you are currently using) as the program to handle sound and files such as MP3s when surfing the web. iTunes can do a pretty good job of dealing with online audio, so in general hitting yes is a good idea, but if you'd rather stick with your existing Internet audio setup, hit No.

• Do you want to search for music files on your computer and copy them to the iTunes Library?
If you have music files scattered around your computer, and you'd like them automatically put in one place, select Yes and iTunes will find and import them all. Otherwise, hit No, as this option can cause random

sound files from the depths of your computer to be imported – you can always remove them, of course, but it's usually nicer to start with a blank sheet and import only the files you actually want.

Getting ready to sync – with and without wires

Once everything's up and running you'll be presented with the iTunes window. To the left is the Source List, which contains icons for everything from playlists to connected iPhones. Click any item in the Source List to reveal its contents in the main section of the window.

Source List

iPhone
icon

By default, you'll only see your iPhone in iTunes when it's physically connected, but most users will want to enable wireless syncing: this way you can sync whenever your iPhone and iTunes are both connected to your home network. To do this, connect your phone to your computer, click its icon in iTunes and under the Summary tab, check Sync this iPhone over Wi-Fi.

Synchronizing

Whenever your iPhone is showing up in iTunes, clicking its icon will reveal various tabs that control how the phone is synchronized with the computer. These include the following, many of which are covered in more detail elsewhere in this book.

• **Info** Lets you synchronize contacts, calendars and bookmarks – though this is usually best left to iCloud. You can also sync mail accounts (i.e. login details and preferences), though it's normally just as easy to log in to your mail on the iPhone directly.

• **Apps** Lets you browse all the apps downloaded via your phone or through iTunes. You can choose which ones to sync over to your phone and even rearrange your apps into iPhone screens and folders – which is often quicker than doing it on the phone itself.

• **Music, podcasts, video & TV shows** Lets you choose which of your iTunes content to sync over to your iPhone. If you'd rather simply drag and drop music and video onto your iPhone, rather than have it synchronized, click Summary > Manually Manage Music and Videos. Note that media downloaded directly onto the phone, or playlists and track ratings created on the move, are copied to iTunes when you sync.

• **Photos** iTunes moves photos from your selected application or folder (see p.138) and gives you the option, each time you connect, of importing photos taken with the iPhone's camera onto your computer.

• **Books** This is the tab where you choose how to synchronize PDF and eBook files (the latter in the ePub format) to be read on the iPhone within the iBooks app. To find out more, turn to p.193.

Whenever your phone is connected, you can click the triangle to the left of its icon in iTunes to see what music and other media it's currently storing. You can also click the triangle to the right of its icon to "eject it" and stop it displaying.

Forcing a sync

When you choose from any of the above options, click Apply Now to start syncing straight away. You can also initiate a sync at any time by right-clicking the icon for your iPhone (or Ctrl-clicking on a Mac) and choosing Sync from the dropdown menu.

Syncing with iTunes on multiple computers

When you connect your iPhone to a different computer, it will appear in iTunes (as long as it's a recent version of iTunes) with all the sync options unchecked. You can then skip through the various tabs and choose to overwrite some or all of the current content.

• **Music, video & podcasts** Adding music, video or podcasts from a second computer will erase all of the existing media from the phone, since an iPhone can be linked with only one iTunes Library at a time. This applies even if you have Manually Manage Music and Videos turned on. You'll also lose any on-the-go playlists and track ratings entered since your last sync. Next time you connect at home, you can reload your own media, but you won't be able to copy the new material back onto your computer.

• **Apps** You can add and use apps from a second Mac or PC (even if it uses a different Apple ID) without overwriting existing apps on the iPhone. However, you will need the password with which the apps were purchased.

• **Photos** can be synced from a new machine without affecting music, video or any other content. However, to leave everything other than photos intact, you need to hit Cancel when iTunes offers to sync the "Account Information" from the new machine.

• **Info** When you add contacts, calendars, email accounts or bookmarks from a second computer, you have two choices – either merging the new and existing data, or simply overwriting the existing data. iTunes will ask you which way you want to play it when you check the box for a category and click Apply. However, you can bypass this by scrolling down to the bottom of the Info panel and checking the relevant overwrite boxes.

> ➜ **TIP** You can use multiple Apple ID accounts to make purchases on an iPhone, but those purchases will only play on iTunes if the computer is "authorized" for the account used (see p.169).

Basic setup options

Once your iPhone is stocked up with all your media and data, you're ready to make it your own.

Ringtones and alerts

To audition the built-in ringtones, tap Settings > Sounds > Ringtones. Here you can also toggle various alert sounds on and off, and choose a volume level for your chosen ringtone (this does the same as using the volume buttons on the side of the phone). If you don't find anything you like, you could download or make a custom ringtone – see box below.

Ring, vibrate and silent

The iPhone offers two modes – Ring and Silent – which you can select using the switch on the left-hand side of the phone. You can choose to have the phone vibrate in one or both of these modes. Click Settings > Sounds to make your selection.

Custom ringtones

If the ringtones that come pre-installed on the iPhone aren't to your taste, try adding your own custom ringtones – which can be done in various ways. The method Apple would prefer you to use is to buy custom ringtones of your favourite songs from the iTunes Store. This is easily done by tapping iTunes > More > Tones – or by searching iTunes and looking under the Ringtones header in the results. However, only a fraction of the songs in the store are available as ringtones, and even if you've already purchased the track you have to pay all over again to get the ringtone.

There are, however, various other ways to get custom ringtones on your iPhone. One option is to use an app – such as Ringtones – that will let you choose from a library of sounds or quickly design your own ringtones by choosing extracts from songs that are already in your phone.

It's also possible to make custom ringtones on a Mac by using Apple's GarageBand software. First, select a loop of under 40 seconds (using the Cycle button) from either your own composition or an existing song file that you've imported. Then choose Send Ringtone to iTunes… from the Share menu, and the ringtone will appear in iTunes ready to sync over to your iPhone.

Wallpaper

Tap Settings > Wallpaper to choose the photos that appear on your Lock Screen (the one you see when you wake the iPhone from sleep mode) and your Home Screen (the screen on which all your icons sit). You can drag and crop the photo before setting it. When setting a Lock Screen image, set the top and bottom edges of a horizontal photo to align with the edges of the semi-transparent zones; the pic will appear framed by, rather than behind, the strips when the phone awakes from sleep. The Home Screen image needs a little more consideration, as a fussy image can make your iPhone pretty ugly and hard to use. The plainer the better … or you could try something custom, such as that pictured here (found at tinyurl.com/iPh4Wall), which mimics the iBooks app bookshelf.

iPhone name

You will probably also want to change the unimaginative name that iTunes gives your iPhone during activation. To do this, tap on the name next to the iPhone icon in iTunes and retype whatever you want.

Auto-Lock

Within Settings > General, you can set the number of minutes of inactivity before your iPhone goes to sleep and locks its screen. In order to maximize battery life, leave it set to one minute unless you find this setting annoying.

Passcode Lock

If you want to protect the private data on your phone – and make sure no one ever uses it to make calls without your permission – apply a passcode. Tap Settings > General > Passcode Lock, and choose a 4-digit number. With this feature turned on, the phone can't be unlocked after waking from sleep without first entering the number. If you forget the code, connect your phone to your computer and restore it from a backup (see p.259). If a thief tries this, they'll get the phone working, but only after blanking all of your private data during the restore process. If you would rather use a proper alphanumeric password instead of a 4-digit pin, slide the Simple Passcode setting off and enter something more complicated.

Note that, by default, a password lock won't stop access to Siri (see p.53). So even though a thief wouldn't be able to access your phone properly, they could still ask Siri to – for example – look up a phone number or address. If this is a concern, switch off the Siri option on the Passcode settings page.

> **→ TIP** Another potential privacy issue is that, even when the screen is locked, the iPhone displays information – such as new SMS messages – as notifications. If you'd rather this didn't happen, tap Settings > Notifications and switch off View in Lock Screen for Messages and any other apps you're concerned about.

In the same page, you can also choose to turn on the Erase Data setting, so that after ten failed passcode attempts, your iPhone wipes itself.

Parental controls

The iPhone also features parental controls, branded Restrictions. Locate the relevant panel within Settings > General and you'll find a host of ways to make the phone more child-friendly – blocking access to YouTube, for example, or to iTunes Store and App Store content tagged as "explicit". If you wish, you can also use this panel to ban app purchases, location services or even use of the iPhone's camera – but you'd have to be a pretty mean parent to turn them all off.

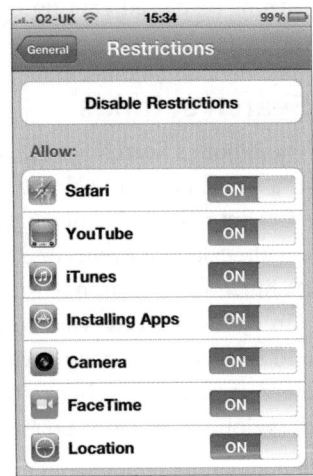

Accessibility tools

The iPhone comes with several accessibility tools designed to help people whose impaired vision, hearing or motor skills make it hard for them to use the iPhone. To see what's available, and play with the options, tap through to Settings > General > Accessibility. The wide range of features include:

• **VoiceOver** Used to speak items of text on the screen. Touch an item to hear it spoken, double-tap to select it and use three fingers to scroll.

• **Zoom** Once set up, a double-tap with three fingers zooms in and out; while zoomed, three-finger dragging moves you around the screen.

• **AssistiveTouch** Allows you to "pre-record" finger gestures – such as pinch zooms – and then play them back using a floating menu. This is useful for people with limited finger dexterity.

• **Audio settings** Aside from enabling mono audio for the headphone socket of your iPhone, you can also set the phone to automatically speak auto-corrections and auto-capitalizations.

• **Video settings** For those whose hearing is impaired, an option for turning on closed-captioning for video can be found within Settings > iPod.

Search settings

The iPhone's Search Screen can be accessed by either a swipe to the left, or a single tap of the Home button.

Get into the habit of quickly popping to the Search Screen to find emails, calendar events or a friend's contact details. You'll soon find that it's much faster than navigating via specific apps.

To choose exactly what content from your iPhone is searched, check and uncheck the options within Settings > General > Spotlight Search. Here, you can also change the order in which results are displayed, by dragging the ≡ icons up and down.

Notifications and the Notification Center

One of the best features to arrive with iOS 5 was the Notification Center, which is accessed by simply dragging down from the very top of the screen in either orientation. What you'll see depends on how you set it up, but typically the list of notifications would include all kinds of recent activity, from calendar alerts and unread emails through to app-specific items such as newflashes or Twitter mentions.

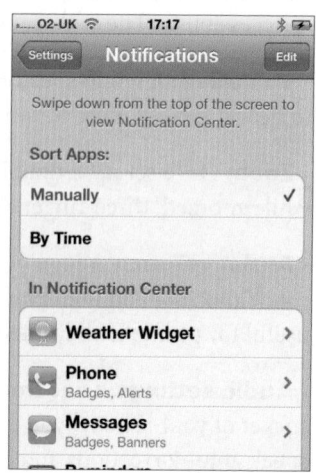

To customize what you see in your Notification Center, tap Settings > Notifications and tweak the settings for each app in turn. In the same screens, you can also decide how you'd like pop-up alerts from each app to appear – e.g. with or without an icon, visible or not on the lock screen, and as a narrow banner or a larger alert with a dismiss button.

04

Typing & Siri

How to get the best from the keyboard and voice controls

The iPhone's touch-screen keyboard isn't to everyone's taste, but once you get used to it it's possible to type surprisingly fast. Even more impressive is Siri – the voice control software introduced with the iPhone 4S. Following are some tips to get you started with both.

Typing

A good place to practise typing is in the built-in Notes app. Open it up and try the following techniques.

The basics

• **Numbers and punctuation** To reveal these keys, tap **?123**.

• **Symbols** To reveal the symbol keys, tap **?123** followed by **#+=**.

• **Moving the cursor** You can tap anywhere in your text to jump to that point. For more accuracy, tap, hold and then slide around to see a magnifying glass containing a cursor.

• **Pop-up keys** The iPhone enters a letter or symbol when you release your finger, not when you first touch the screen. So if you're struggling to type accurately (or you're entering a password and can't see what you're typing) try tapping and holding the letter. If the wrong letter pops up, slide to the correct one and then release.

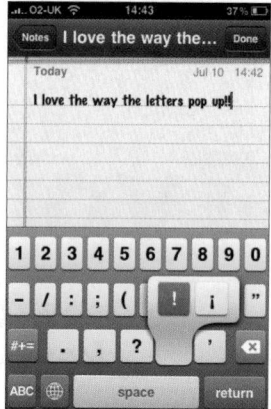

> **→ TIP** Tapping and holding on a letter is also the way to reach special characters and accents. For example, tap and hold down on "e" and you'll see é, è and other options.

Speed-typing tips

• **Quick full stops** When you reach the end of a sentence, double-tap the space bar to add a full stop and a space. If this trick doesn't work, turn the relevant option on in Settings > General > Keyboard.

• **One-touch punctuation and caps** Tap 123 and then slide to the relevant key without taking your finger off the screen. The same trick works with capital letters: tap shift and slide to a letter.

• **Caps lock** If you like to TYPE IN CAPS, turn on caps lock by double-tapping the Shift key (⇧). If it doesn't work, first enable this feature under Settings > General > Keyboard.

• **Thumbs and fingers** Apple recommend that you start off using just your index finger and progress to two thumbs. However, you could also try multiple fingers on one hand. This can be even faster, since thumbs tend to bump into each other when aiming at keys near the middle of the keyboard, though it does take a little getting used to.

Auto-correct and spell check

It's tricky to hit every key accurately on the iPhone, but usually that doesn't matter much, thanks to the device's word-prediction software. Even if you hit only half the right letters, the phone will usually work out what you meant by looking at the keys adjacent to the ones you tapped and comparing each permutation of letters to those in its dictionary of words. However, no such system is perfect and the iPhone often gets it wrong – sometimes with hilarious results, as sites such as damnyouautocorrect.com attest.

Accepting and rejecting suggestions

When the iPhone suggests a word or name it will appear in a little bubble under the word you're typing. To accept the suggestion, just keep typing as normal – hit space, return or a punctuation mark. To reject it, finish typing the word and then tap the suggestion bubble before continuing.

Dictionary

The iPhone has a much bigger and more relevant dictionary than most mobile phones – including, for example, many names and swear words. In addition, it learns all names stored in your contacts and any word that you've typed twice and for which you've rejected the suggested correction. Unfortunately, it's not currently possible to edit the dictionary, but you can blank it and start again. Tap Settings > General > Reset > Reset Keyboard Dictionary.

If you're ever stuck for the spelling or meaning of an obscure word, try a dictionary web app such as idotg.com/apps, or one of the many downloadable dictionary apps available in the App Store.

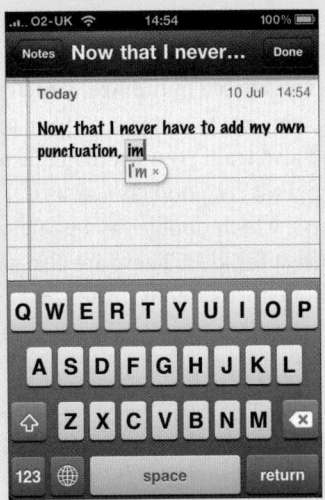

Spell check

After you've typed a word that the iPhone doesn't recognize, it will be underlined in red to suggest that it might be a spelling error. If you want to check, tap the word to be presented with an alternative spelling from the dictionary.

Auto-capitalization

In addition to correcting letters, the iPhone will add punctuation (changing "Im" to "I'm", for example) and capitalize the first letter of words at the start of sentences. If you prefer to stick with lower case, turn off Auto-capitalization within Settings > General > Keyboard.

- **One touch at a time** If typing with two thumbs or multiple fingers, only let one finger touch the screen at a time. If the first finger is still on the screen, the second tap won't be recognized.

- **Landscape keyboard** In Safari, Mail and many other applications, you can rotate the iPhone to get a bigger version of the keyboard.

- **Keyboard shortcuts** One handy recent typing feature is the ability to create shortcuts for words or phrases that you use regularly. For example, you could specify an email address – such as johnsmith@gmail.com – and the iPhone would then automatically offer up the address when you typed the first few letters. To add a word or phrase, tap Settings > General > Keyboard > Add New Shortcut. It's also possible to give each word or phrase a special shortcut, which the iPhone will automatically substitute for the word of phrase each time you type it. For example, you could set it to replace "jsa" with "John Smith Associates".

- **Test your skills** There are loads of apps available in the App Store that you can use to test your typing speed, and also train yourself to get faster. TypeFast is a good choice, as is Typing Class, which doubles as a game, and is great for kids. There are also several iPhone formatted websites that can help, such as iphone.typingweb.com.

→ **TIP** If you fancy something a bit old school, try a virtual old-fashioned typewriter app, such as miTypewriter.

Cut, copy & paste

In terms of typing, among the most useful features on the iPhone are the Cut, Copy and Paste commands – especially if you don't particularly get on with the device's keyboard. Though at first they can seem a little fiddly to use, it really is worth getting acquainted with the taps and drags that bring these controls to life. And we are not only talking about text; these options will also pop up for copying images, and even entire webpages. For now, however, here's the deal with text:

• **Locate the text or word** you want to copy or cut and then tap and hold until an options bubble appears; choose Select to highlight that word, or Select All for the whole piece of prose of which your word is a part. To speed things up, you could also try tapping once at the start of the text you want to select and then immediately dragging to the end of the last word.

• **Use the blue end-point markers** to resize the selection you have highlighted, and then when you are ready, tap Copy or Cut in the option bubble. Both actions move the selection onto the iPhone's "clipboard", ready to be pasted elsewhere.

> ➔ **TIP** If you want a specific app for organizing copied content on a clipboard, download the excellent Pastebot.

• **Navigate** to the place where you want to paste the text (which might be a completely different app, an email you are composing or perhaps the search field in Safari).

• **Tap and hold** at your desired point and choose Paste from the option bubble that appears. You can, alternatively, Select text (as above), then Paste over the top of it.

> **→ TIP** When copying, pasting and typing, you can undo your last action by shaking the iPhone.

Other pop-up typing options

Besides cut, copy and paste, various other tools may pop up when you select text, depending on the app you're using. Tap the little triangle icons, where present, to see all the available tools. They include:

• **Formatting** Really useful when composing emails, you can add bold, italic and underline, as well as adjusting the quote level of each paragraph.

• **Suggest** Presents similar words from the dictionary – useful if you've accidentally ended up with the wrong word thanks to a dodgy autocorrect.

• **Define** Searches for the current word in the iPhone's dictionary.

Siri & Voice Control

The iPhone's voice control software is either brilliant – if you're using an iPhone 4S or some later model – or not much use, if you have an older phone. That's because the 4S, with its faster processor, is the first iPhone to come with Siri – a breakthrough piece of voice control software that works so well that it can be quite unnerving when you first try it.

Activating Siri

There are three main ways to activate Siri:

• **Press the Home button** and keep it held down for a second or two. (This also works for the older Voice Control software.)

• **Raise the phone to your ear**. This works whenever the phone is switched on (not in standby mode) and not engaged on a call.

• **Hold down the control button** on the iPhone's earbuds.

Once the Siri screen appears, you'll hear a chime to tell you to start speaking. Once you're done, you should hear another, higher chime, which signifies that whatever you said is now being processed. At any time you can also press the onscreen purple microphone icon to start or stop a voice instruction.

> **→ TIP** In addition to launching Siri as described above, you can use speech to enter text wherever you would normally type. Just press the microphone icon next to the space bar on the iPhone's virtual keyboard.

Using Siri

The remarkable thing about Siri is that – depending on your accent – it will generally understand what you say even if you make no effort to simplify your speech. However, it is easy to increase the accuracy by, for example, leaving tiny

pauses between words. Softening or avoiding particularly strong regional twangs or turns of phrase will also help. Here are some other things you need to know for getting the most out of Siri:

• **Controlling your phone** Siri can activate almost any kind of activity on your iPhone. For example, try things like "Take a note", "Call John's Mobile", "Play me some Beatles", "Set an alarm for 7am", "Email Anna saying I'm running late".

• **Ask it anything** In addition to controlling your phone, Siri is surprisingly good at responding to factual questions such as "When was the Battle of Waterloo?" or "What's the capital of Ethiopia?" If it can't answer a question, it should offer to google it for you. For fun, see how it responds to questions such as "What's the meaning of life?"

• **Speak your punctuation** When dictating text, you can include punctuation verbally. Virtually any punctuation mark will work – from "comma" and "full stop" (which works better than the American "period") to "exclamation mark", "semi-colon", "dash", "asterisk" and even (for typography pedants) "en-dash" and "em-dash". You can also capitalize a word by adding "all cap" before it.

• **Correcting** When dictating text into an app, such as Notes, aurally ambiguous words and phrases, such as to/too and there/their, are underlined in blue. Click on any underlined text to reveal the alternatives.

05

Connecting

Wi-Fi, Bluetooth and other airwaves

The iPhone can handle various kinds of wireless signal: GSM, for phone calls; GPRS, EDGE and 3G for mobile Internet access; Wi-Fi for Internet access in homes, offices and public hotspots; and Bluetooth for connecting to compatible headsets and carphone systems. This chapter takes a quick look at each of them.

Airplane mode

Airplane mode, quickly accessible at the top of the Settings menu, lets you temporarily disable GSM, GPRS, EDGE, 3G, Wi-Fi and Bluetooth, enabling you to use non-wireless features such as the Music app during a flight or in any other circumstances where mobile phone use is not permitted. (Whether phones actually cause any risk on aircraft is disputed, but that's another story.) Airplane mode can also be useful if you want to make sure that you don't incur roaming charges when overseas by inadvertently checking your mail and so on.

	15:53	100%
Settings		
Airplane Mode	ON	
Wi-Fi	Off >	
Notifications	On >	
Sounds	>	
Brightness	>	
Wallpaper	>	
General	>	
Mail, Contacts, Calendars	>	

Using Wi-Fi

Connecting to networks

To manually connect to a Wi-Fi network, tap Settings > Wi-Fi and choose a network from the list. If it's a secure wireless network (as indicated by the 🔒 icon), the iPhone will invite you to enter the relevant password.

Though this procedure doesn't take long, it's best to have the iPhone point you in the direction of Wi-Fi networks automatically. This way, whenever you open an Internet-based tool such as Maps or Mail, and there are no known networks in range, the iPhone will automatically present you with a list of all the networks it can find. You can turn this feature on and off via Settings > Wi-Fi > Ask to Join Networks.

If the network you want to connect to isn't in the list, you could be out of range, or it could be that it's a "hidden" network, in which case tap Wi-Fi > Other and enter its name, password and password type.

> ➜ **TIP** If you want to maximize battery life, get into the habit of turning Wi-Fi off when you're not using it. It only takes a couple of seconds to turn it back on when you need it again.

Forgetting networks

Once you've connected to a Wi-Fi network, your iPhone will remember it as a trusted network and connect to it automatically whenever you're in range. This is useful, though can be annoying – if, for example, it keeps connecting to a network you once chose accidentally, or one

Finding public hotspots

Many cafés, hotels, airports and other public places offer wireless Internet access, though often you'll have to pay for the privilege of using them – particularly in establishments that are part of big chains. Typically, you connect and sign up onscreen. If you use these kinds of services a lot, you may save time and money by signing up with a service such as BT Openzone, Boingo, T-Mobile and AT&T.

BT Openzone www.btopenzone.com
Boingo boingo.com
T-Mobile t-mobile.com/hotspot
AT&T att.com

The ideal, of course, is to stick to free hotspots. Try the Wi-Fi Finder app (pictured left) to help you locate them. Alternatively, browse an online directory such as:

Hotspot Locations hotspot-locations.com
WiFinder wifinder.com
Wi-Fi Free Spot wififreespot.com

which lets you connect but doesn't provide web access. In these cases, click on the ◉ icon next to the relevant network name and tap Forget This Network. This won't stop you connecting to it manually in the future.

When it won't connect...

If your iPhone refuses to connect to a Wi-Fi network, try again, in case you mistyped the password or tapped the wrong network name. If you still have no luck, try the following:

• **Try WEP Hex** If there's a ◉ icon in the password box, tap it, choose WEP Hex and try again.

• **Check the settings** Some networks, especially in offices, require you to manually enter information such as an IP address. Ask your network administrator for the details and plug them in by clicking ◉ next to the relevant network name.

• **Add your MAC address** Some routers in homes and offices (but not in public hotspots) will only allow access to devices specified in the router's "access list". If this is the case, you'll need to enter the phone's MAC address – which you'll find within Settings > General > About > Wi-Fi Address – to your router's list. This usually means entering the router's setup screen and looking for something titled MAC Filtering or Access List.

• **Reboot the router** If you're at home, try rebooting your wireless router by turning it off or unplugging it for a few seconds. Turn off the Wi-Fi on the phone (Settings > Wi-Fi) until the router has rebooted.

• **Tweak your router settings** If the above doesn't work, try temporarily turning off your router's wireless password to see whether that fixes the problem. If it does, try choosing a different type of password (WEP rather than WPA, for example). If that doesn't help, you could try updating the firmware (internal software) of the router, in case the current version isn't compatible with the iPhone's hardware. Check the manufacturer's website to see if a firmware update is available.

GSM, GPRS, EDGE & 3G

In your home country, the iPhone should automatically connect to your carrier's GSM network for voice calls, and to the fastest data network available – either GPRS, EDGE or 3G. All three networks will automatically give way to Wi-Fi (which is usually much faster) whenever possible.

> **→ TIP** If you have an iPhone with 3G and want to maximize battery life, consider turning the 3G connection off within Settings > Network. Once that's done, your phone will use EDGE instead, which is slower but far less power-hungry.

Test your speed

To test the speed and latency (time lag) of your current EDGE, 3G or Wi-Fi signal, try a free app such as Speedtest, which you can use to keep records of your connection speeds from day to day; alternatively, tap Safari and visit:

iPhone Speed Test testmyiphone.com

Connecting abroad

When overseas, voice calls should work normally, though North Americans may first need to activate international roaming with their carrier. You'll probably find that your phone selects foreign carriers automatically. If you prefer, however, it's possible to specify a preference. Simply tap Settings > Carrier and pick from the list. For more on making calls abroad, see p.19.

As for the Internet, Wi-Fi should work wherever you are in the world – for free, if you can find a non-charging hotspot. For mobile Internet, GPRS, EDGE or 3G will work if there's a compatible network and you've turned on Data Roaming within Settings > General > Network. But be prepared for some extremely steep usage charges (see p.19).

Bluetooth

Bluetooth allows computers, phones, printers and other devices to communicate at high speeds over short distances. The iPhone uses Bluetooth to share an internet connection with a computer or iPad (see overleaf) or to connect to headsets and certain carphone systems. You can also connect an iPhone 4S to a full-size Bluetooth keyboard (see p.273) – useful if you've got a lot of typing to do.

You can turn Bluetooth on and off by tapping Settings > General > Bluetooth. If you're not using it, leave it switched off to help maximize battery life.

Personal Hotspot – "tethering"

Although not everyone realizes it, it's possible to use an iPhone 3G or any later model to provide wireless Internet access to one or more other devices, such as laptops and iPads. This can be enormously useful when you're out and about with a laptop or iPad and unable to connect online except via your phone. Here's how it works: the iPhone gets online in the normal way via 3G or EDGE and then share that internet connection with the laptops or iPads either via Wi-Fi (up to five devices), Bluetooth (up to three devices) or USB (one device). The only downside is that some mobile phone networks don't support "tethering" – as this function is traditionally known – and others charge extra for it.

If your carrier and tariff do support tethering, you can turn it on within Settings > General > Network > Personal Hotspot. Here you can also choose a password for the Wi-Fi network that will automatically be established by your phone. Once that's all done, check your computer and you should see a Wi-Fi network with your iPhone's name. Select this, enter the password you chose and you should now be online. Alternatively, connect the computer and phone with a USB cable or pair them using Bluetooth.

Connecting to office networks

Exchange – for calendars, contacts, etc

Most offices run their email, contacts and calendars via a system known as Microsoft Exchange Server. Individual workers access these tools using Outlook, and a web address is usually made available to allow remote log in from home or elsewhere. If necessary, it's usually possible to access your work account via the web on the iPhone – just press Safari and go to the regular remote-access web address. However, it's much neater to point your phone at the Microsoft Exchange Servers directly – see p.118 to find out more.

VPN access

A VPN, or virtual private network, allows a private office network to be made available over the Internet. If your office network uses a VPN to allow access to an Intranet, file servers or whatever, you'll probably find that your iPhone can connect to it. The phone supports most VPN systems (specifically, those which use L2TP or PPTP protocols), so ask your administrator for details and enter them under Settings > General > Network > VPN.

Remote access

As well as connecting via a VPN, it's also possible for the iPhone to connect directly to a computer that's on the Internet – a Mac or PC in your home or office, say. Once set up, you could, for example, browse your files or stream music and video from your iTunes collection. All you need is the right app and the correct security settings set up on the computer in question.

Some of the various apps available for this purpose include NetPortal (pictured), LogMeIn Ignition and TeamViewer. However, Mac owners with an Apple AirPort router should look no further than Back to My Mac, which is included as part of iCloud.

06
Apps

Downloading & organizing iPhone applications

Perhaps the single best thing about the iPhone is the sheer number of applications – or apps – available for it. There are tens of thousands of apps out there, the vast majority of which can be downloaded inexpensively or for free. An app might be anything from a driving game to a metro map, from a tool for making calls across the Internet to a version of your favourite website optimized to work perfectly on the iPhone's small screen. If you can imagine an application for your iPhone, someone has probably already created it and made it available.

The iPhone comes pre-installed with a bunch of apps – from Mail and Safari to Weather and Stocks. But these are only the tip of the iceberg. To see what else is available, dive into the App Store…

Downloading apps

Just as the iTunes Music Store changed music-buying habits, Apple's App Store – which provides apps for the iPhone, iPod Touch and iPad – is quickly changing the way in which software is distributed.

With so many apps available, the main problem is the potential for getting somewhat addicted and spending more time and money than you meant to.

To use the App Store you'll need your Apple ID username and password (see p.68).

Accessing the App Store

The App Store can be accessed in two ways:

• **On the iPhone** Simply click the App Store icon. Assuming you're online, you can either browse by category, search for something you know (or hope) exists, or take a look at what's new, popular or featured. When you find something you want, hit its price tag (or the word "free") and follow the prompts to set it downloading. You'll need to enter the Apple ID that you already set up when you activated your iPhone.

Although apps can be downloaded via 3G or EDGE, larger ones can take an eternity with anything less than a Wi-Fi connection.

• **On a Mac or PC** Open iTunes, click the Store icon and then hit App Store in the top menu. All the same apps are available and the interface for browsing them is, if anything, better than the one on the phone itself. Any apps you download in iTunes will be copied across to your iPhone next time you sync. Or, if you have one Apple ID for iTunes and iCloud, you can access apps bought on any device by opening the App Store on your phone and looking in the Purchased list.

Note that there are no refunds in the App Store, so it pays to read reviews before you buy.

> → **TIP** Be careful if buying via iTunes that you choose apps that are suitable for the iPhone – some are iPad-only. Apps suitable for both devices are known as universal and marked with a **+** sign.

Buying apps with multiple accounts

Neither the iPhone, nor iTunes on your computer, have to be wedded to a single Apple ID. So if more than one member of your household uses the same iPhone, iPad, Mac or PC, there's no reason why you can't all have your own IDs and buy apps separately. Once installed on the phone, all the apps will be available to use, whichever account is currently logged in to the iPhone's Store. However, to update an app, you'll need the password for the account through which it was purchased.

> → **TIP** To log out of your account on the iPhone, either tap Settings > Store > Sign Out, or scroll to the bottom of most App Store windows and tap your Account: name button.

Updating apps

One of the best features of the App Store is that as and when developers release updates for their software, you will automatically be informed of the update and given the option to install it for free, even if you had to shell out for the original download. To update apps:

• **On the iPhone** The number of available updates is displayed within a red badge on the corner of the App Store's icon. Tap App Store > Updates and then either tap Update All or choose individual apps to update one at a time.

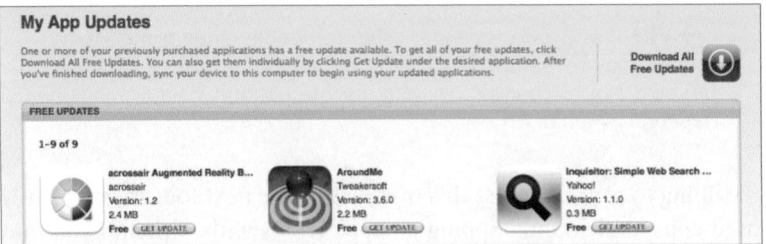

• **On a Mac or PC** The number of available updates is displayed next to the Apps header in the sidebar of iTunes. Click Apps followed by Check for updates, bottom-right, to see what's available.

> **→ TIP** In Settings > Store you can decide if your Newsstand apps (see p.205) will automatically update by downloading new content automatically.

Running and switching apps

To launch an app on your iPhone, simply tap its icon. To close it, press the iPhone's Home button – though as we'll see, the app may continue to run in the background.

Fast app switching

If your iPhone has iOS 4 or later, then double-clicking the Home button when your phone is in use reveals a panel showing your recently used applications. This is extremely handy, enabling you to quickly switch between apps without having to return to the Home Screen.

→ **TIP** If you own an older iPhone, double-tapping the Home button may reveal the Music controls or your Favorites contacts. if you'd like to change this, look for the option within Settings > General > Home.

Swiping to the left across this tray reveals the next four most recently used apps, and so on. Tapping an app's icon swiftly switches you into that app, and for most apps you'll find that you are back in exactly the same place you were last time you used it.

Should you want to remove apps from the recently used panel to restart them properly, tap and hold the icons until they all start to jiggle about; then tap the red ⊖ icon of the apps you want rid of. Click the Home button when you are done. This process does not remove the app from the iPhone all together: only from this list.

→ **TIP** The iPhone allows multitasking, which means apps can perform tasks in the background while you get on with something else in another app. However, Apple has limited the types of features that can run in the background in order to preserve battery life and stop the phone grinding to a halt. Background functions supported include music playback, receiving VoIP calls, location awareness and various others.

Switching to the Music app

Swiping to the right across the recently used app tray reveals music controls and an icon that gives you quick access to the Music app. To the left of these playback controls there's a special button for locking the iPhone's screen orientation to portrait – handy when trying to use reading or web-browsing apps in bed. When the screen orientation is locked, this button displays a padlock in its centre.

Web apps & webclips

Downloading apps is one way to fill up your Home Screen with icons; the other is to create simple bookmark icons – so-called "webclips" – that point to your favourite websites or web apps (see box overleaf).

These can be handy as they allow you to access your most frequently viewed websites without opening up Safari and manually tapping in an address or searching through your bookmarks.

To create an icon for a website, simply visit the page in Safari and press ✚ (or the ⬆ icon on some older iOS versions). Select "Add to Home Screen" and choose a name for the icon – the shorter the better, as anything longer than around ten characters won't display in full.

Improving website icons

When you add a webpage to your Home Screen, the iPhone will use that website's iPhone icon, if one has been specified. If the site doesn't have an iPhone icon, an icon will be created based on how the page was being displayed when you clicked ✚. To make the best-looking and most readable icons, zoom in first on the logo of the website in question.

If you're not happy with the icon you get, it is possible – if a bit of a hassle – to create your own icons for those websites. The process involves creating an icon (or grabbing one from elsewhere), uploading it to the web, and then using a special bookmark to point the iPhone to this icon before you hit ✚. To grab the special bookmark and get more details, see:

AllInTheHead.com tinyurl.com/addiphoneicons

Web apps and optimized sites

Whereas a "proper" iPhone app is downloaded to your phone and runs as a standalone piece of software, a "web app" is an application that takes the form of an interactive webpage. The term is also often used to describe plain old webpages specially designed to fit on the iPhone screen without the need for zooming in and out. As such, you might also hear people refer to them simply as iPhone-optimized webpages and websites.

Unlike many "proper" apps, web apps will usually work only when you're online. On the plus side, they tend to be free – and they usually take up no space in your iPhone's memory. There are exceptions to these rules, however. For example, Google's Gmail web app uses a "database" or "super cookie" (see p.215) to store some information locally – which does take up space and enables some functions to be carried out even when the iPhone is offline.

Several websites – including Apple's own – offer directories of iPhone-optimized websites and web apps, arranged into categories, though increasingly these sites look out of date, due to the rise of fully fledged iPhone apps.

Apple's web app directory
apple.com/webapps

Whether or not they call them "web apps", many popular websites offer a version for browsing on mobile devices in general or iPhones specifically. These not only fit nicely on small screens but also load much faster, since they come without the large graphics and other bells and whistles often found on the full version of their websites.

Usually, if the server recognizes that you're using a mobile web browser, the mobile version of a site will appear automatically. Occasionally, though , you'll have to navigate to the mobile version manually – look for a link on the homepage. Mobile sites often have the same address but with "m", "mobile" or "iphone" after a slash or in place of the "www". For example:

Digg digg.com/iphone
eBay mobile.ebay.co.uk
Facebook iphone.facebook.com
Flickr m.flickr.com
MySpace m.myspace.com

Organizing your apps

As your iPhone fills up with apps and webclips, they will soon start to spill over onto multiple screens, which you can easily switch between with the flick of a finger – simply slide left or right.

The number of screens currently in use is shown by the row of dots along the bottom of the screen, with the dot representing the current screen highlighted in white. You can return to the Home Screen by pressing the Home button or by sliding left until you reach it.

Moving icons

To rearrange the apps and webclips on your iPhone, simply touch any icon for a few seconds until all the icons start to wobble. You can now drag any icon into a new position, including onto the "Dock" at the bottom of the screen (to put a new icon here, first drag one of the existing four out of the way to clear a space).

To drag an icon onto a different screen, drag it to the right- or left-hand side of the screen. (To create a new screen, simply drag an icon to the right-hand edge of the last

Checking app settings

Just like the apps that come pre-installed on the iPhone, many third-party apps have various preferences and settings available. Many people overlook such options and can end up missing out on certain features as a result. Each app does its own thing, but expect to find some settings either:

• **In the app** If there is nothing obvious labelled Options or Settings, look for a cog icon, or perhaps something buried within a More menu.

• **In iPhone Settings** Tap Settings on the Home Screen and scroll down to see if your app has a listing in the lower section of the screen. Tap it to see what options are available.

existing screen.) Once everything is laid out how you want it, click the Home button to fix it all in place.

App folders

Instead of just having your apps and links arranged across various screens, it's also possible to group relevant icons together into folders. You might create, for example, a "travel" folder, which contains a mix of map-based apps, guidebooks and links to travel-related websites.

You could simply group related icons together onto specific screens, but folders offer some advantages. For example, they can be titled and live on your Home Screen, which makes them instantly available at all times.

To create a folder, first touch any icon for a few seconds until they all start wobbling. Then simply drag one icon onto another, and a folder will be created containing both icons. The iPhone may attempt to give your new folder a name based on the categories of apps you've combined – tap into the title field to overwrite this with a name of your choice. To add another app to the folder, simply drag it onto the folder's icon. To remove one, click on the folder and simply drag the relevant icon out.

> ➡ **TIP** Once you have lots of icons, you may find it easier to organize them into screens and folders in iTunes, which allows you to see multiple screens at once and to move icons around with a mouse. To do this, connect your iPhone to your computer, select its icon in iTunes and click the Apps tab.

Deleting apps

It's worth noting that your Apple ID account keeps a permanent record of which apps you have downloaded, so if you do delete both your iPhone's copy and the iTunes copy, you can go back to the Store and download it at no extra charge.

If you delete a webclip, you can reinstate it by restoring from a backup, though it is generally easier just to go back to the website in question and make a fresh one (see 73).

You can't delete any of the built-in apps, but you can move them out of the way onto a separate screen or into a separate folder if you don't use them.

• **Deleting apps and webclips from the iPhone** Hold down any Home Screen icon until they all start jiggling and then tap the small red ⊗ of the app you want to delete. When you have finished, hit the iPhone's Home button. This will not remove the app from iTunes, so you can always sync it back to your iPhone again later.

Phone

07

Calls

From conference calling to video chat

Though the iPhone is special in many ways, it largely sticks to the familiar when it comes to making and receiving calls. This chapter whizzes you through the basics, offering some useful tips and tricks along the way.

Making calls

There are various different ways of making a call using the iPhone. Simply tap the Phone button, and then…

• Tap Contacts, then a name, then a number.

• Tap Keypad and type a number (hit ⊗ if you make a mistake). Then tap Call. You can even paste in a number that you've copied in another app by tapping the space just above the keypad and then hitting Paste.

• Tap Favorites and hit a name. Because each Favorite is a specific number, this can save you a couple of taps.

• To call someone you recently phoned, or who recently phoned you, click Recents and hit the relevant name or number.

You can also launch calls from outside of the Phone app. For example:

• Use voice-activated dialling (see p.93).

• Click on a number in an email, webpage, message or elsewhere.

• Tap on a missed or recent call in Notification Centre by dragging down from the top of the screen.

> → **TIP** Even a locked iPhone can still call emergency services. Tap Emergency Call on the lock screen and dial, for example, 911 in the US or 999 in the UK, then tap Call.

When you're on a call...

When the iPhone is close to your ear, a proximity sensor disables the screen, saving you power. Move the phone away from your ear while on a call and the screen displays various call options:

• **Mute** When tapped, the caller can't hear you. Tap again to return to normal. Hold down Mute for a couple of seconds and it will switch onto Hold until you tap it again.

• **Add Call** For multiple and conference calls (see overleaf).

• **Keypad** Brings up the numeric keypad
– essential when using automated phone
systems.

• **FaceTime** Tap to initiate a video call
with another iPhone 4, 4S or iPad 2 or
Mac user – see p.91.

• **Speaker** Toggles speaker-phone on
and off. The iPhone's speaker is located on the base of the unit. As with
other phones, the sound isn't amazing and may distort, in which case
reduce the volume level using the buttons on the side of the iPhone.

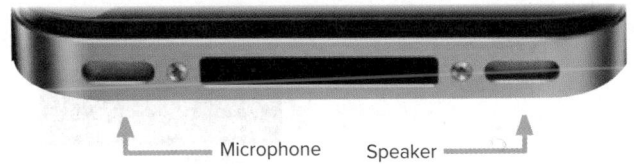

Microphone   Speaker

• **Contacts** Takes you to your main Contacts list.

And more...

While on a call, you can press the Home button to access any other
applications without dropping your call – useful if you need to check
a date in your calendar or an address in an email. Note, however, that
you can only access the Internet during a call when you're connected to
a Wi-Fi network (see p.62). So don't expect to use Safari or Maps while
walking down the street in the middle of a conversation.

When you have finished with whatever it was you were doing, press
the Home button again and then tap the green strip at the top of the
screen to return to the Call Options screen.

Multiple calls and conference calling

The iPhone has two lines available for calls. When you're talking to one person, you can click Add Call and dial someone else; the existing caller will be put on hold, and you can then use the Swap option to switch between the two lines. Likewise, if someone calls you while you're on the phone, you'll be offered a Hold Call + Answer option.

Even better, the iPhone allows you to make conference calls with up to five people simultaneously. With a conference call, you don't need to switch between callers – everyone can be heard by everyone else. It works like this: one of the iPhone's two lines hosts the conference and the other is free to call people who can then be merged onto the conference line.

So, to get things started, make or receive a call in the usual way. Next, tap Add Call and dial someone else. The first call is put on hold while you do this so you can give the second person you've called warning that you want to add them to a conference – or, indeed, chat privately. Then, hit Merge Calls to create the three-way conversation. Follow the same procedure to add further callers to the party.

First caller (on hold) ——————▶

Second caller (live) ——————▶

Merge calls to start
a conference ——————▶

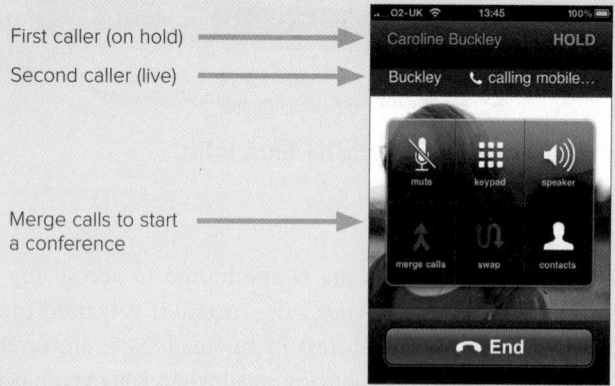

During a conference call, you can also...

• **Talk privately** with a particular caller. Tap Conference and then Private next to the relevant name. To bring both of you back to the conference, tap Merge.

• **Ditch a caller** Tap Conference, then tap End next to the relevant name.

• **Add an incoming call** Tap Hold Call + Answer and then Merge Calls.

Receiving calls

When someone calls you, the iPhone will either ring or vibrate (for setting up ringtones, see p.48) and display the caller's information on the screen, including their photograph, if you have one set up in Contacts (see p.137). Next, do one of the following:

To answer a call

To talk when a call comes in, tap Answer or, if the iPhone is locked, drag the slider. If, when the call comes in, you have audio or video playing, it will fade out and pause. If you're using the supplied headset, click the headset's mic button to answer the call.

To decline a call

If you don't want to talk, declining a call will send it straight to voicemail. This can be achieved either by tapping Decline on the screen, or:

• Pressing the Sleep/Wake button on the top of the iPhone twice in quick succession.

• If you are using the iPhone headset, press and hold the mic button for a couple of seconds, then let go. You will then hear two low beeps confirming that the call has been declined.

To silence a call

When a call comes in, you can quickly stop the iPhone ringing or vibrating without answering or declining the call – useful if, for instance, you get a personal call in the office and you want to step outside before answering it.

To do this, simply press the Sleep/Wake button or either of the volume buttons.

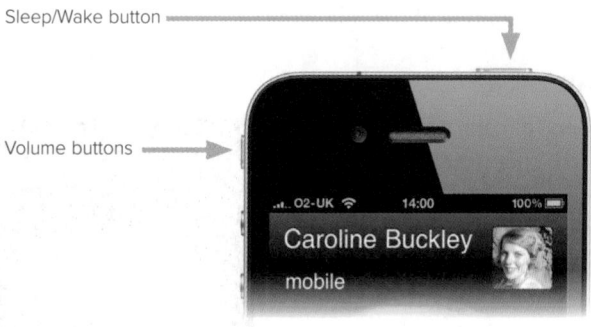

Sleep/Wake button

Volume buttons

To answer a second call

If you have Call Waiting switched on (in Settings > Phone), you can receive a second call while on the phone. The iPhone will chirp in your ear, show the new caller information and offer you three options:

• **Ignore** Sends the new caller to voicemail.

• **End Call + Answer** Ends the call you were on and answers the new one.

• **Hold Call + Answer** Puts the first call on hold and answers the second. From there you can either switch between the two conversations using the Swap button or hit Merge to combine the calls into a three-way conference.

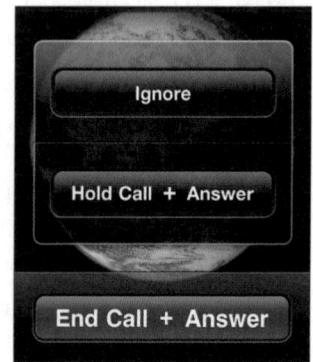

Recent and missed calls

Like all mobile phones, the iPhone keeps a list of recent incoming and outgoing calls. This can be used for reference (showing you the time when someone phoned, say) or as a means of storing numbers or making calls. To access the list, tap Phone > Recents. If there's a red circle with a number in it on the Recents icon, that's telling you the number of missed calls listed since you last looked.

In the list, missed calls appear in red and can be viewed in isolation by tapping the Missed button. When a caller has attempted to reach you more than once, the number of missed calls is displayed in brackets.

Tapping ⊚ to the right of any entry will display more information about the call, such as whether it was incoming or outgoing. When a caller is already in your Contacts list, all their information is displayed, with the number that relates to that specific call highlighted in blue.

Call notifications

By default, missed calls can be accessed in the Notification Centre. To view the list, simply drag down from the top of the iPhone's screen. To call someone shown in the list, simply tap on the relevant name or number.

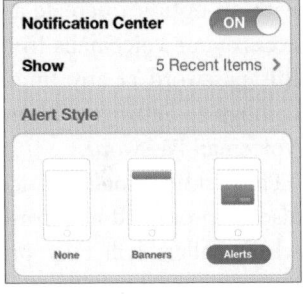

You can turn this feature on and off – and select the number of calls included in the list – by tapping Settings > Notifications > Phone. The same screen reveals various other options, too. For example, you can choose whether you'd like missed calls to show as alerts (which appear in the centre of the screen and need to be dismissed) or banners (which appear at the top of the screen and can disappear without instruction). You can also switch off Show In Lock Screen if you'd like to ensure that no one with physical access to the phone can see the list of your missed calls.

Visual Voicemail

One of the most innovative features ushered in by the original iPhone (though still not available on all networks) was Visual Voicemail. The idea is that you no longer have to listen to all of your voicemail messages in turn in order to get to the one you want. Instead, your voicemails are presented as a list – just like emails – and you can choose the ones you want to listen to in any order. You can even rewind and fast-forward. Use the system for just a few weeks and you will hardly believe it was ever done the old way.

As mentioned, the Phone button on the Home Screen displays a red circle containing the total number of missed calls and unheard voicemails. Tapping it reveals the Voicemail button, which also displays a red numbered icon, but this time just for unheard voicemails.

Voicemail setup

Tapping the Voicemail button for the first time takes you to a screen with an option to create a voicemail password, which you can use to access your voicemails from other phones (see p.90); you can change

this password at any time by tapping through to Settings > Phone > Change Voicemail Password.

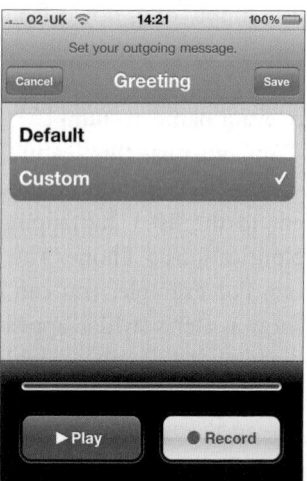

The first time you tap Voicemail, you'll also be prompted to record a greeting, which callers will hear prior to leaving you a message. Tap Greeting, then Custom, then Record. When you're done, you can play back your message and, if you're happy with it, tap Save. Alternatively, if you're feeling shy, you could tap Default instead of Custom and stick with the pre-recorded greeting, which includes your number.

Voicemail overseas

In some foreign countries, you may find that Visual Voicemail won't work. Instead, when you click Voicemail, you'll be offered a single Call Voicemail button which will take you through to your messages the old-fashioned way. As with other phones, you can also call your voicemail by holding down the 1 on the numeric keypad.

By default, the iPhone will alert you with a sound when you have a new voicemail (except if the Silent switch is on). If you'd rather turn this function off, tap Settings > Sounds and set the New Voicemail switch to Off.

Playback and more

The Voicemail screen lists current voicemail messages, with those you have not yet listened to displaying a blue dot to their left. It is worth knowing that even though a voicemail appears in the list, it hasn't been downloaded to the phone. Thus, you need to have a network signal to listen to voicemails.

Each voicemail lets you…

• **Play/pause/rewind** You can play and pause a message at any time with the ▶ and ❚❚ icons. Even better, you can rewind or skip forward by dragging the scrubber bar: one of the iPhone's finest moments in terms of innovation.

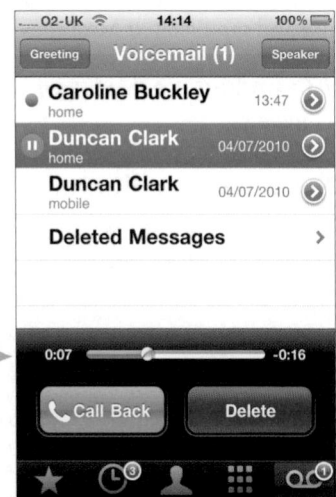

Scrubber bar

• **Return a call** Tap the message and hit Call Back.

• **View details** Tap the ❯ button to the right of any message to find out the time and date it was recorded, its duration and the full contact info of the caller, where known.

• **Contact the caller** Having tapped ❯, you can tap the caller's number to call them, their email address to send an email, or send an SMS, etc.

• **Add to Contacts** Tap ❯ next to a message and then Create New Contact, Add to Existing Contact or Add to Favorites.

Deleting and undeleting voicemails

One great thing about the iPhone's voicemail system is that deleted messages are saved for thirty days before being permanently erased.

• **To delete a message** Tap the message, then tap Delete.

• **To view deleted messages** Scroll to the bottom of the voicemail list and tap Deleted Messages.

• **To undelete a message** Choose a deleted message and click Undelete.

Picking up voicemail from another phone

iPhone owners in many countries can pick up their voicemails the old-fashioned way using any phone. Simply call your iPhone's number

Voicemail over speaker and Bluetooth

To listen to your voicemail messages over the iPhone's built-in speaker, tap the Speaker button in the top-right corner. Or, if your iPhone is connected to a Bluetooth headset or car kit (see p.270), tap Audio and choose Speaker Phone to use the built-in speaker. To switch back to the headset or car kit, tap Audio again, then choose the relevant device.

and, assuming it is not answered, you'll be redirected to your voicemail. When you hear your greeting, dial * followed by your voicemail password (see p.50). Then enter # and follow the voice instructions.

FaceTime

Easily the most important new feature of the fourth-generation iPhone was FaceTime, a foolproof way of making and receiving video calls. Unimaginable outside science fiction just a decade ago, mobile video calls offer a whole new type of communication – the downside being that it's only possible to make or receive FaceTime calls using a recent iPhone (4 and later), iPod Touch (4 and later), iPad (2 and later) or Mac (OS X 10.6.6 or later).

Video calling required the addition of an extra camera lens on the front of the iPhone so that the user's face can be captured without spinning the phone around and obscuring the screen. Best of all, because

there's a camera lens on the back too, it's possible at the tap of a button to switch between the two views – one of you and one of your view.

Using FaceTime

Assuming FaceTime is switched on (you can check within Settings > Phone), then making a video call is as easy as tapping a button – though you'll need to be on either Wi-Fi or 3G to do it (see p.62). To initiate a video call, either:

• Tap the FaceTime button on a contact's page in your contacts list.

• Make or receive a phone call in the usual way and then hit the FaceTime button on the call screen (pictured here).

Either way, the person you are calling will immediately receive a FaceTime invitation. If they accept, the video call will begin. During the call, each user can toggle between their iPhone's two cameras using the "switch camera" icon located at the bottom-right of the screen that displays while you are using FaceTime.

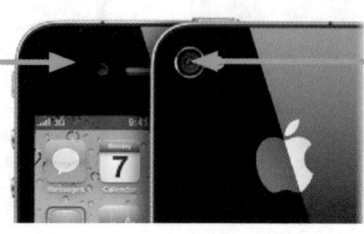

Front lens
for showing the
caller's face

Rear lens
for showing
whatever
the caller is
looking at

Using FaceTime on Macs

FaceTime is also available as a 99¢/69p download for Macs, enabling you to call friends, family and colleagues on their FaceTime-capable iPhones (4 and later models). You can also call Mac to Mac, if both computers have cameras and FaceTime installed – a similar experience to using Skype or iChat.

To use FaceTime, first download it from the Mac App Store. To make a call, choose a contact from the list. Then, to call that person's iPhone, click their phone number; to call their iPad, iPod Touch or Mac, click their email address.

Voice-activated dialling

On any iPhone since the 3G model, it's possible to initiate a call using your voice – though it works best with Siri (see p.53), which was introduced on the iPhone 4S.

• **To call a contact** Press and hold the Home button for around a second until the Voice Control screen pops up and you hear a double-beep. Then say "call" or "dial" followed by a contact's name. You can add "home" or "mobile" if you know the contact has multiple numbers. If you don't do this, your iPhone will next ask you to clarify which number you want.

• **To call a number** Press and hold the Home button until you hear a noise. Then say "call" or "dial", followed by the number, speaking each digit. On older iPhones, avoid phrases such as "double" and "triple", which can cause confusion.

Other call features

Call Forwarding

If you'd like to have incoming calls forwarded to another number, click Settings > Phone > Call Forwarding and enter a number. This can be very handy if, for instance, you're going to be outside your network's coverage area but available on a landline. When this feature is active, a special blue icon appears at the top of the iPhone's display to the left of the clock.

Caller ID (outgoing)

If you'd like to call someone without your name or number flashing up on their phone screen, click Settings, then Phone, and switch off Show My Caller ID.

Super Caller ID

Tap ◉ by a received call in your Recents list and you'll see options for calling back, adding to contacts, etc. This page includes address details, when known, for callers already in your contacts list. For unknown callers, if you're in the US, you'll see the area where the call came from. This "Super Caller ID" feature can be very handy when you receive a call with an unfamiliar area code. Unfortunately, the same info doesn't flash up when the phone actually rings.

08

Apps: calls

From four-way video chat to free international calls

There are hundreds of apps available to augment your iPhone calling experience. Many offer pure novelty, while others bring interesting and genuinely useful extra functions.

Fun calling apps

Dial Plate
This is one of many apps that add a retro rotary dialler to your iPhone's armoury. Pictured right.

FakeCaller
A fun way to schedule "fake" incoming calls. Useful if you want to impress your friend by having the President phone up – or if you need an excuse to get away from a bad social function.

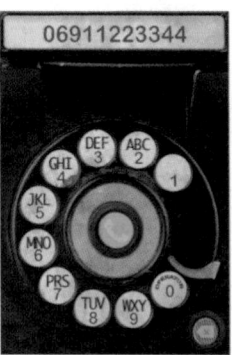

FaceDialer

A handy app for adding individual numbers to your Home Screen with a photo icon of the contact. There are dozens of similar apps avail-

able that add icons to the Home Screen for shortcut calling friends and loved ones. Call Him and Call Her are among the more stylish offerings for speedy spouse dialling.

Calling via the Internet

Anyone accustomed to using Skype or other Internet calling software on their computer will be aware that it's possible to make free or virtually free Internet calls – voice and video – to landlines and mobiles all over the world. Given that the iPhone is essentially a small computer that can connect to the Internet, it's not surprising that it allows you to do just the same. Here are some of the options.

Skype

Easily the most popular Internet calling system, Skype is available for free

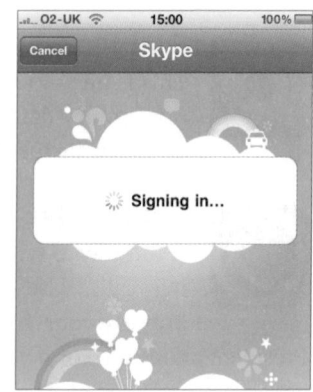

from the App Store. Once it's installed, either log in with your existing Skype account details or set up a new account and you'll immediately have access to all the same features familiar from the computer version of Skype – instant messaging, Skype-to-Skype calls and, if you buy some credit, very inexpensive

Skype-to-phone calls. Skype-to-Skype calls can be made for free via Wi-Fi and 3G – though bear in mind that you'll be chomping through your data allowance when using 3G.

Truphone

A very good (and free) alternative to the Skype app is Truphone. It gives you VoIP call contact with Skype, Truphone and Google Talk users and also a free voicemail service with push notifications. It works really well over Wi-Fi and 3G and can run in the background, allowing you to receive calls from the service at any time.

Fring

Like Truphone, Fring gives you access to contacts on a range of networks, this time including Windows Live Messenger, Google Talk, Twitter, Yahoo, AIM and ICQ. The service's main offering, however, is its ability to support four-way video calls – a bit like a conference call version of FaceTime.

Jajah

Although apps such as Skype and Truphone are the most obvious way to make Internet calls via your iPhone, there are other options too. Sign up for a free Jajah account, for example, and you can call other Jajah users for free – even internationally – and make other long-distance calls at discounted rates (as little as 3¢/1.5p for the US, China and most of Western Europe). You go to the website using Safari and enter your own number and the number you want to call.

Press Go and your phone will ring; pick it up and the number you're calling will ring. Neither of you initiated the call, so neither of you will pay long-distance charges.

Jajah iphone.jajah.com

 ## Rebtel

Rebtel works slightly differently. Enter the phone number of a friend, relative or colleague in any of the fifty or so supported countries and Rebtel will rank them into "free" and "cheap". For cheap, fees are as little as 1¢/1p per minute.

The system also works on landline and on non-app mobiles: you go to the website and enter the phone number of a friend, relative or colleague in any of the fifty or so supported countries and Rebtel will give you a local number on which you can call using your free minutes, plus a tiny fee to Rebtel. Best of all, the numbers stick, so you can save one to your phone and use it whenever you want to call the same person.

Rebtel rebtel.com

→ **TIP** When making calls via an Internet-based app, you'll often need to enter numbers complete with international dialling codes, even if you're not calling overseas. So it's good to get into the habit of using dialling codes when adding or editing contacts.

Calling apps for Mac and PC

One advantage of being able to easily sync all your phone numbers from your phone to your computer is that you can then import them into an Internet telephony app and call the numbers from your computer in order to save your precious mobile minutes. For example, the free desktop version of Skype, which is free to download, lets you import numbers from Address Book or Outlook. Just choose Import Contacts… in the Contacts menu.

Skype skype.com (PC & Mac)

To maximize call quality from your computer, try headphones, an audio headset or USB handset. It is possible to get by with the microphone and speakers built into most laptops, but using headphones or a handset improves the quality by eliminating any feedback and echoes caused by the sound from the speakers getting picked up by the microphone.

More calling apps

calLog
When used in place of the iPhone's built-in call list interface, this app allows you to add notes and reminders to your calls and also to export reports on call activity.

0870
Mobile phone users in the UK get stung by high calling rates to 0870 and 0845 numbers. This app provides a conventional landline substitute for each number.

Record Phone Calls
One of various apps offering the ability to record calls. Handy, though for ethical (and sometimes legal) reasons, it's important to tell the other caller that they're being recorded.

ChatTime
A service that allows inexpensive international calls via the regular phone network – rather than calling via the Internet.

09
Contacts

Importing, syncing & managing

The iPhone is a world-class digital address book, providing space not only for names and numbers but for all sorts of other information – from job title to Skype name. Even better, the iPhone, with the help of iTunes and iCloud, makes it easy to synchronize your contacts with the address book on your computer or iPad. This is not only convenient, it also means that your precious contact details can be safely backed up.

If you've been living on the moon for the last decade, you might wish to start your contacts list from scratch, in which case turn to p.106 to read about using the Contacts app on the iPhone. Probably, though, you'll want to import some contacts from elsewhere. Contacts that are already stored in the Address Book on your Mac or PC – or in a Google or Yahoo! account – can be synced easily via iTunes (see p.35). And corporate contacts can usually be imported along with a corporate email account (see p.118). If you need to import contacts from an old phone, it can be a bit trickier, but the next few pages should help.

Importing contacts from an old phone

There are various ways to import your numbers from an old phone, as described below. However, if you only have a few numbers to move across, or the following techniques seem like too much hassle, you could just manually enter your numbers into your computer (see p.103) or iPhone (see p.106).

Option 1: Use a SIM card

See if your old phone has an option to move all contacts to the SIM card. Next, insert that SIM into your iPhone and go to Settings > Email, Contacts, Calendar > SIM Contacts. If the SIM is too big for the iPhone, or the iPhone doesn't recognize it, you can either have it chopped down to microSIM size at a mobile phone shop, or use a SIM reader (see box below) to import the contacts directly onto your Mac or PC and drop the contacts into the address book that you sync with your phone.

SIM readers

A SIM reader is an inexpensive device that will feed numbers (though not pictures, etc) stored on a mobile phone SIM card into a Mac or PC via a USB socket. Before using one, you'll need to make sure that the numbers on the phone are stored on the SIM rather than in the phone's memory. Consult the manual to find out how this is done (you can probably download the manual if you no longer have it). Then remove the SIM, insert it into the reader and connect it to your computer. You'll probably be left with vCard or text files, which should be easy to import into Address Book, Outlook or Outlook Express (look for the Import option in the File menu).

If you're unlucky, and you can't get the files to import, try opening them with Text Edit (Mac) or Notepad (PC). You should then at least be able to see the data from your phone and, if needs be, copy and paste it into new contact cards.

Option 2: Let the old phone talk to your computer

Many mobile phones can connect to a Mac or PC, using either Bluetooth or a cable. If your phone came with a cable, try that first. Otherwise, make sure your old phone is set to be "discoverable" (an option usually found under Connectivity or Bluetooth) and try to connect it to your computer using Bluetooth. On a Mac this is done by clicking > System Preferences > Bluetooth > Devices > Set Up New Device. On a PC look within the control panel.

Once the phone and computer are connected, the next challenge is to import the contacts. Check to see if the manufacturer of your old phone offers a downloadable contact import tool. Otherwise, try Googling for a third-party tool that can speak to your particular phone model. These tend to cost $20–40 or so but will help you copy across photos and movies, as well as contact info. Examples include:

DataPilot datapilot.com
Mobile Master mobile-master.com
WinFonie Mobile 2 bertels-hirsch.de/en/winfonie_mobile_2

Alternatively, if you have access to a Mac running OS X 10.6 or earlier (check within > About This Mac), you could try using the iSync tool that you'll find in the Applications folder, which may allow you to import the contacts onto a new group on that computer's Address Book – which in turn can be exported and sent to your own computer.

If you manage to extract the contacts, in most cases the resulting file can be easily dropped into Outlook or the Mac or Windows Address Book (choose Import in the File menu), ready to sync to your iPhone.

Option 3: Ask a phone shop for help

Many mobile phone stores will pull the numbers off an old phone and provide them via email or CD, ready to be imported into your contact application of choice. Try your carrier first as they have a free service. Otherwise, try an independent store, though be prepared to have to pay a small fee.

Editing contacts on a computer

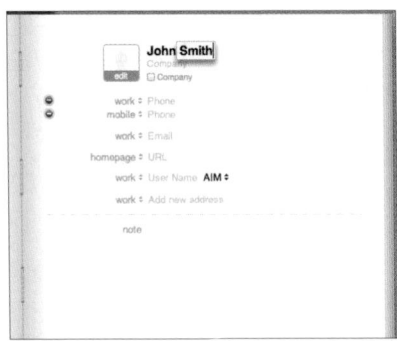

However you get the numbers from your old phone onto your computer, it's likely that the result will need some editing – duplicates culled, email addresses and other information added, and so on. All this can be done on the iPhone (as we'll see), but it's easier on a computer.

It's beyond the scope of this book to describe contact management in Address Book or Outlook in any depth, but here are some general pointers:

• **Default formats** Look within your Preferences or Options to check that the default address and phone number format is correct for the country where you live.

• **Groups** You can add contacts in your address book to "groups" or "distribution lists". You might have all your work colleagues in one group, say, and all your friends and family in another. When you sync your iPhone you can choose to copy across either all contacts or just selected groups.

• **Duplicate entries** After you import contacts from an old phone, you may end up with duplicated entries. The Mac Address Book has a "Merge Contact" feature that can resolve the problem. Select the two entries when holding down the ⌘ key and choose Merge Selected Cards from the Cards menu. Outlook, meanwhile, will automatically alert you when duplicate entries appear and ask you how you want to resolve the conflict. To check if this feature is enabled, open Options from the Tools menu.

• **Pictures** Though time-consuming and not particularly useful, adding images to contacts can be a fun way to personalize your Address Book on your computer and, in turn, your iPhone. It can be especially useful if you are the kind of person who finds it hard to put names to faces. You can, of course, take snaps out and about with the iPhone itself and associate those images with your contacts then and there (see p.137), but pictures can also be added to your computer's address book. On a Mac, simply drag a photo onto an entry.

Syncing contacts with the iPhone

If you're a Mac user and you've got your contacts in Address Book, then getting them onto your iPhone is a cinch. Just turn on iCloud and enable contacts syncing on each device, as described on p.41. Once that's done, a contact added on the iPhone should automatically show up on your other iCloud-enabled devices – and vice versa.

However, iCloud won't work if your contacts are stored on a PC; if you're using an older iPhone model that can't connect to iCloud; or if your contacts are all stored in Gmail or Yahoo! rather than Address Book. In these cases, you'll need to sync your contacts the old-fashioned way using iTunes. To do this, open iTunes, connect your iPhone and click its icon. Switch on contacts syncing under the "Info" tab and choose whether you want to import all contacts or just particular groups.

Syncing contacts with Entourage

Apple describe Entourage – the Mac version of Outlook – as being compatible with the iPhone in terms of syncing contacts and calendars. But this isn't possible directly. In order to enable syncing between the iPhone and Entourage, you first have to enable syncing between Entourage and the Apple Address Book. To do this, open Entourage and choose Preferences from the Entourage menu. Under Sync Services, check Contacts – and Events if you want to sync calendars.

Contacts on the iPhone

Browsing contacts

You can access your full list of contacts either from within the main Phone screen or by opening the separate Address Book app. To browse the list, either flick up or down with your finger, or drag your finger over the alphabetic list on the right to quickly navigate to a specific letter. Alternatively, tap the search box at the top of the screen and type in the first few characters of the name you're looking for.

When you find the contact you want, tap once to view all their details, and then tap a phone number to start a call or an email address to launch a new mail message.

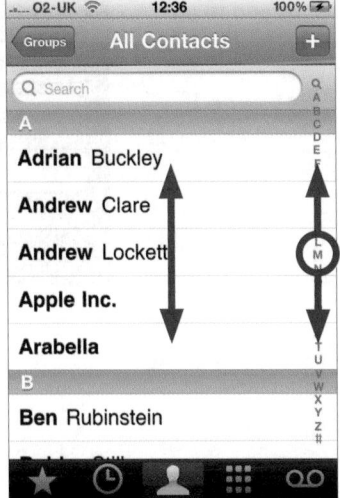

Flick up and down to home in on a specific contact

Tap once to stop a moving list in its place

Drag up and down the alphabet to speed to a specific letter

→ TIP By default the iPhone alphabetizes your contacts by surname. You can change this in Settings > Mail, Contacts, Calendar.

Adding contacts

There are various ways to add new contacts on the iPhone.

• **In Contacts** When viewing your contacts list, click the **+** icon.

• **From the keypad** You can enter a number via the keypad (within Phone) and then tap the head-and-shoulders graphic to create a new contact or add the number to an existing contact.

• **From the recent calls list** When viewing recent calls (under Phone), you can tap the ❯ icon next to an unrecognized number and choose to create a new contact or add to an existing contact.

• **From an email or webpage** If someone emails you a number, or you stumble across someone's number online, tap and hold it until the options to either Create New Contact or Add To Existing Contact appear.

> **→ TIP** You can also add contacts to your list by photographing business cards using a downloadable app such as BC Reader.

Favorites

The Favorites list, found within the main Phone screen, provides quick access to your most frequently dialled numbers. Instead of full contacts, it stores a specific number for each name. This way you can dial with just a couple of clicks – much quicker than browsing through a long contacts list, selecting a person and then picking a number.

• **To add Favorites** Browse your contacts and click the relevant names, in each case choosing Add to Favorites and picking a number. Alternatively, from Favorites, hit the + button to browse for names.

• **To change the order** of your Favorites, tap the Edit button and drag contacts by the ☰ icon to their right.

• **To remove a Favorite** Click Edit, tap the red ⊖ icon next to the relevant name, and then hit Delete.

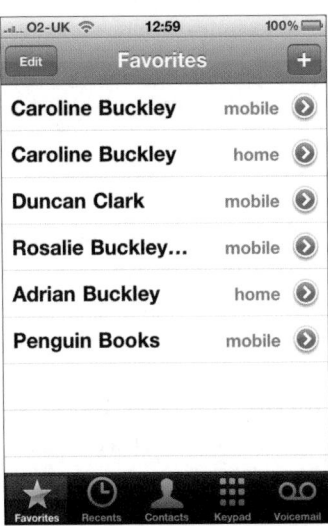

> → **TIP** When a number is added as a Favorite, a small star appears next to the relevant number on the contact's page.

Editing contacts

To edit a contact – change their name or number, assign a specific ringtone, add an email address, or whatever – first select a name in the main contacts list and hit Edit. Then...

• **To add a new number, address or other attribute to a contact**, tap the green ⊕ icon.

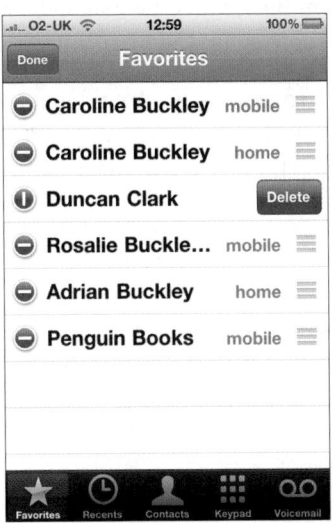

• **If you can't see the relevant attribute**, click Add Field. (The same command can be found on a Mac within iCal's Card menu.) You'll then be offered everything from "birthday" to a space for notes. Birthdays is particularly useful, as you can have those dates show up automatically in the Calendar app. If you can't see them there, tap Calendars and switch on Birthdays Calendar. On a Mac, you can also have your birthdays show up in iCal by clicking Preferences > Show Birthdays Calendar.

• **To delete an item**, tap its red ⊖ icon. To delete the contact from your Contacts list entirely, scroll down to the bottom of the entry and tap the "Delete Contact" button.

• **To assign a picture to a contact**, click Add Photo (top left) and choose to either take a photo with the iPhone's camera or to select an existing photo already stored on the phone. Once you've selected an image, you can crop it by pinching and dragging. When you're done, click Set Photo. Alternatively, start by selecting a picture in Photos or Camera, hit the arrow button and tap Assign to Contact.

10

Messages

All about texting: SMS & iMessage

"Texting", also known as text messaging or SMS, is familiar to anyone who has ever used a mobile phone. However, on the iPhone, it works a little differently. For a start, you get a full **QWERTY** keyboard (no more multi-tapping number keys) and your texts are shown as threaded conversations. The second difference is that you're not limited to old-fashioned **SMS** text or picture messages. When sending a message to an iPhone or iPad running iOS 5, you can send a free **iMessage** instead.

The basics: SMS & iMessage

The iPhone has always offered standard text messaging, and for a number of years has enabled users to send photo and video SMS messages too. But in late 2011, the iOS 5 software ushered in a third type of message: the iMessage. It works like this: you open the Messages app and write and send a message as usual. If you're sending it to a normal phone, it will be sent as an SMS. But if you're sending it to another iPhone running iOS 5 or later (or to an email address of someone using iMessage on an iPad or iPod Touch) the app will automatically send an iMessage instead – assuming you're online, that is. iMessages offer a number of advantages over traditional SMS messages.

• **Zero cost** iMessages are sent via the Internet, rather than through the phone network. This means they're effectively free: you can send as many as you like without worrying about your monthly SMS allocation or fees for sending picture or video messages.

• **Receipts** When you send an iMessage, you'll get a delivery receipt to confirm that it reached the iPhone or iPad you were sending it to. And if the recipient has chosen to switch on read receipts, you'll also get a confirmation that they actually saw your message.

• **Not just phones** An SMS can usually only be sent and received by a mobile phone, but iMessages also work with the iPad and iPod Touch.

Message settings

To decide how your phone handles iMessages and SMS, click Settings > Messages. Here you can choose whether you want to allow read receipts – and, if necessary, completely disable either iMessages, SMS messages or both.

The same settings screen also allows you to check and change the phone number and Apple ID email address linked with your phone's use of iMessages – but this is best left as it is unless something isn't working.

Message notifications

When you have unread messages, the Message icon on the Home Screen will show a small red circle with a digit reflecting the number of new messages. If you have iOS 5, you can also quickly access recently received unread texts by dragging down the notification area (see p.52).

To control whether or not your iPhone plays an alert sound when you receive a message tap Settings > Notifications > and use the Text Tone slider. (The sound won't play if the ringer button is switched to Silent.) You can also change if and how visual message notification appears – handy if, for example, you want to avoid message previews popping up on arrival for privacy reasons. To access these settings, tap Settings > Notifications > Messages.

Using Messages

Clicking the green Messages icon reveals a list of unread messages (signified by a blue dot) and existing "conversations". Tap an entry to view one and you're ready to reply. Alternatively:

• **To delete a message or conversation** Swipe left or right over it to reveal the Delete button. Alternatively, tap Edit, then ⊖.

• **To write a new message** Tap ✐ and either enter a phone number, start typing the name of someone in your Contacts list to reveal matching names, or hit ⊕ to browse for a contact.

> ➜ **TIP** To see whether your message will be sent as a regular text or a free iMessage, look in the text-entry box before you start typing. It should say either "Text Message" or "iMessage".

• **To send a message to multiple people** Start a new message (you can't do this via an existing conversation) and tap ⊕ to add new names. Note that if it's an iMessage, recipients will be able to reply to all. If it's a text message, they will only be able to reply to you.

• **To see if your message was delivered** Look for the Delivered label in small type underneath each message. Note that this only works for iMessages, not regular text messages.

• **To forward a message** tap Edit, then select one or more speech bubbles and choose forward.

> → **TIP** To quickly get to the top of a long SMS conversation and access the Call and Contact Info buttons, tap the time at the top of the iPhone's screen.

• **To quickly send a message to someone in your Favorites or Recents lists** Tap ⦾ next to their name and choose Send Message.

• **To add a photo or video** Tap the camera icon by the text area and either shoot a new pic or video, or choose one already on the phone. (Alternatively, start by finding a video or photo you want to share, and tap the arrow button followed by Send Message.)

• **To call or email someone from your Text Messages list** Tap a message in the list, scroll to the top of the conversation and tap Call or, to see their other numbers and email address, click Contact.

• **To add someone you've already texted as a Contact** Tap their phone number in the Messages list and then tap Add to Contacts.

> → **TIP** Street addresses, emails, weblinks or phone numbers in a text conversation can be tapped to launch Maps, Mail or Safari, or to start a call.

• To see if an outgoing message was send as an iMessage

The iPhone will automatically send iMessages to other iOS 5 users.

However, it only works if both phone or iPads are online. If not, the text will be delivered as an SMS and you'll see a small "Sent as SMS" below the relevant speech bubble in the conversation.

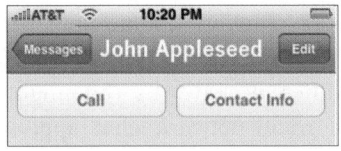

Messaging apps

Emoji

If you want to use emoticons, also known as emoji icons, in your messages and emails, you could download an app for the purpose. There are loads of free and paid-for options available – Emoji by JG Applications being a good choice. It works by adding a special custom keyboard set to your iPhone's armoury. Once the app is installed, navigate to Settings > General > Keyboard > International Keyboards > Japanese and enable the Emoji option. From then on, the new keyboard is available via the keyboard switching "globe" button to the left of the spacebar on the iPhone keyboard.

Trillian, etc

In addition to using Apple's Messages app, you can also use the iPhone to access every other chat and messaging service, such as AIM, Facebook Chat, Google Talk, iChat, ICQ, MSN, Skype and Yahoo! This can be handy for chatting with friends on their computers or non-Apple phones. Although it's possible to use chat services via Safari, it's usually far better to download an app designed for the job. Most of the big chat services offer an iPhone app, though if you have contacts spread across many different networks you might prefer a multi-network chat app, such as Trillian (pictured), BeeJive or IM+, which allow you to chat simultaneously across a range of services.

Texting from your computer

Though we're used to sending text messages from our phones, it's also possible to do so from a computer – which can occasionally be handy. Some websites (including those of many phone carriers) allow you to send free SMS messages, but it's also possible to send texts to US numbers from some chat applications. In AOL Instant Messenger or iChat, for instance, just create a new message from the File menu and enter the recipient's mobile number where you'd usually enter a "screen name". You'll need to include +1 at the beginning, as SMS messages sent through the Internet require a country code.

If you want to text someone in a foreign country, but don't want to pay international rates via your mobile plan, you can use Skype (see p.96) to send texts to most countries for around 8¢/5p. However, you'll want to first change your sender ID to your iPhone number so that people can reply directly. Note that the ID won't show up correctly in the US, China or Taiwan.

Email & calendars

11

Email

How to set up and use Mail

Having email available wherever you are completely changes your relationship with it. It becomes more like text messaging – but much better. As it is with iPads and Macs, the iPhone's email application is known as Mail – and although the iPhone version is much more limited, it has evolved into a great little app.

Setting up email accounts

The iPhone comes pre-configured to work with all the leading email systems – including Yahoo!, Hotmail and, for corporate systems, Microsoft Exchange – without a full setup process. Gmail is also ready-to-go, but for the best possible service you may want to set up your Gmail as an Exchange account using Google Sync (see p.119).

To set up an account the simple way, simply tap Settings > Mail, Contacts, Calendars > Add Account. Choose your account provider from the list and enter your normal login details. Under Description, give your email account a label, for example "Personal" or "Work" – this is your means of distinguishing between multiple email accounts within the iPhone's Mail application.

You may be prompted to log in to your account on the web and enable IMAP or POP3 access (opt for IMAP if given the choice). You can do this via Safari on your iPhone, or using a computer.

> **→ TIP** Corporate email systems are usually based on Microsoft Exchange. This is fully compatible with the iPhone – though it's up to your network administrator whether or not to allow access on iPhones or other devices. Speak to your network administrator if you can't get it to work – and if they're not sure what the problem is, ask if they'd mind enabling IMAP on the server.

Setting up email accounts that aren't in the list

If you use an email account provided by an Internet Service Provider or some other system that's not ready-to-go on the iPhone, there are two ways to get that account up and running on your phone – either copy across the account details from your computer using iTunes, or enter the details directly into the phone.

Using iTunes

iTunes can sync your email account details between your iPhone and Mail or Outlook on your computer. This won't copy across the actual messages – just the login and server details, etc. To get things going, connect your iPhone to your Mac or PC, click its icon in iTunes and choose the Info tab in the main panel. Scroll down, check the box for each account you want to copy across, and press Apply.

Entering the details on the iPhone

To manually set up an email account on the iPhone, tap Settings > Mail, Contacts, Calendars > Add Account... > Other. Then enter all the details for your account. If you're not sure of some of the entries – such as the mail server addresses – try contacting your email provider and asking them. If you'd rather not spend ages in a customer services phone queue, try searching the Internet: most providers have a page on their website that spells out everything you need to know. It's also often possible to guess the details from the email address. If your email address is joebloggs@myisp.com, the username may very well be joebloggs (or your full email address), the incoming server may be mail.myisp.com or pop.myisp.com; and your outgoing server may be smtp.myisp.com.

Contacts and calendars too?

Some email accounts come with extra services such as calendars, contacts and notes. You can turn those on and off selectively for each account within Settings > Mail, Contacts, Calendars.

Push versus fetch

Traditionally, a computer or phone only receives new emails when its mail application contacts the relevant server and checks for new messages. On a computer, this happens automatically every few minutes – and whenever you click the Check Mail or Send/Receive button. This is referred to by Apple as a "fetch" setup.

By contrast, email accounts that support the "push" system feed messages to the iPhone the moment they arrive on the server – which is usually just seconds after your correspondent clicks the Send button. iCloud and Yahoo! both support the push system for emails, contacts and calendars – and so do recent installations of Microsoft Exchange. If you use Gmail, you don't get push services by default at the time of writing – though it is possible to get them by setting up your Gmail account through Exchange using Google Sync:

Google Sync google.com/mobile/sync

You can turn push services on and off within Settings > Mail, Contacts, Calendars > Fetch New Data. Here you can also set up how frequently you would like accounts that use fetch to check for new emails. It's worth noting that the more frequently emails and other data are fetched, the quicker your battery will run down. These settings also apply to other apps (such as some to-do list tools and instant messaging clients) that rely on Apple's push services to grab your up-to-date data from a server.

Using Mail

Using email on the iPhone works just as you'd expect. Tap Mail on the Home Screen to get started. If you have more than one email account set up, you can choose to either view each Inbox separately, or view them all together by tapping the All Inboxes option. You can also…

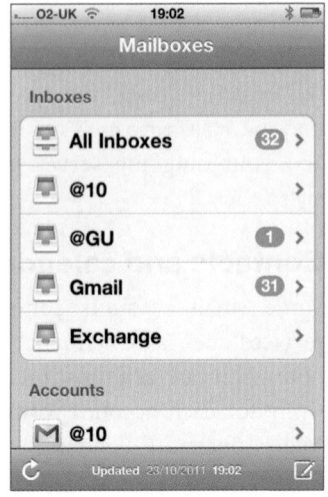

• **Compose a message** Tap ✐. (If you have more than one account set up, first select the account you want to use from the list.) Alternatively, you can kick-start a message using Siri (see p.53) or by tapping a contact's email address in any app.

> ➔ **TIP** Although it's easy to miss, you can add formatting to your emails – including bold, italic and indented quotes. Tap to select some text and then tap ▶ on the popover to reveal the options.

• **View a message** Tap any email listed in your Inbox to view the entire message. Double-tap and "pinch" to zoom in and out – just like with Safari. If you often find that you have to zoom in to read the text, try raising the minimum text size under Settings > Mail.

> ➔ **TIP** Jump to the top of long emails or scrolling lists by tapping the iPhone's Status Bar at the top of the screen.

• **Move between messages** Use the ▲ and ▼ buttons at the top to move up and down through your emails.

• **Move between accounts and folders** Use the left-pointing arrow button at the top (which displays the name of the item one layer up in the hierarchy) to navigate through all your folders and accounts, with the latter farthest to the left.

• **Open an attachment** You can open (though not necessarily edit) Word, Excel, PowerPoint and iWork files attached to emails. You can also view images and PDFs from emails and save them to Photos and iBooks, respectively, to be synced across to your computer. To save an image, tap it and choose Save Photo. To save a PDF tap the Open in iBooks button.

• **Reply or forward** Open a message and tap ↰.

• **Moving messages** To move one or more messages to a different folder, hit Edit, check the messages and tap Move. Or, when viewing an individual message, tap the 📬 button.

• **To attach a photo** You can't add an attachment to a message that you have already started directly from Mail, but you can Paste images that have already been copied using the Copy command in another application. To reveal the Paste command, tap and hold within the message you are composing. You can also create an email from images in the Photos app: select an image, tap 📨, and follow the

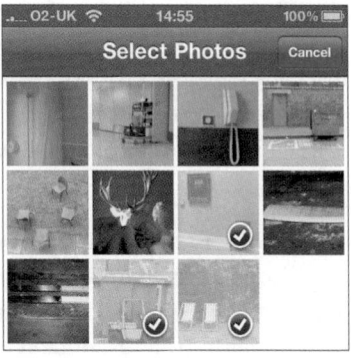

prompts. To send multiple images from Photos while viewing a grid of images (either an Album, Places or Faces set), tap 📨, then select the images you want to attach to an email and then hit the Share button. If you have multiple email accounts, messages that originate from Photos will be sent from the default account, which you can select from Settings > Mail, Contacts, Calendars.

• **Deleting and archiving messages** You can delete a message from a list by swiping left or right over it and then tapping Delete. To delete multiple messages simultaneously, tap Edit and check each of the messages you want to trash. Then click the Delete button. Note that iCloud and Gmail accounts may offer the option to "archive" rather than delete messages – though you can change this if you like (look within the Advanced Settings of a given account).

> **→ TIP** A third way to delete messages is to open them and press 🗑. This then jumps to the next message. If you don't want to waste time confirming each time you hit delete, turn off the Settings > Mail > Ask Before Deleting option.

• **Empty the Trash** Each email account offers a Trash folder alongside the Inbox, Drafts and Sent folders. When viewing the contents of the Trash, you can tap Edit and either permanently delete individual items or choose to Delete All. Alternatively, tap Settings > Mail, Contacts, Calendars, choose an account, and then tap Account Info > Advanced > Remove to set how often messages in the Trash are automatically deleted – either never, or after a day, a week or a month.

• **Create new contacts from an email** The iPhone automatically recognizes phone numbers, as well as email and postal addresses when they appear in an email. Simply press and hold the relevant text to see the options to Add to Contacts or Copy to the clipboard. In the case of a postal address, you will also get the option to see the location in Google Maps.

> **→ TIP** As with Safari, you can tap and hold a link in an email to reveal the full destination address. Useful for checking the location of links before you click on them.

Tweaking the settings

Once your email account is up and running on your phone, scan through the Settings options to see what suits you. Tap Preferences > Mail, Contacts, Calendar and explore the following under the Mail heading. Some of the most useful options include…

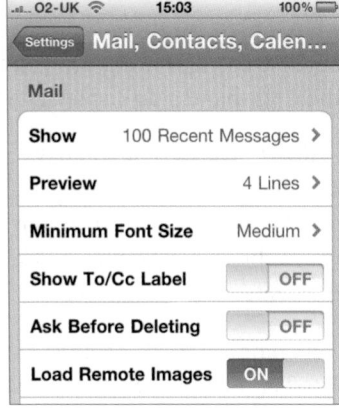

• **Show** Use this option to determine how many messages are displayed within your Inbox – useful if you are in denial about the amount of mail you have to get through!

• **Preview** Lets you set the number of lines of text that will be visible in your mailboxes for each message before you open it.

• **Show To/Cc Label** Lets you see at a glance whether you were included in the To: or Cc: field of an email. When switched on, a small icon will appear by each message preview stating "to" or "cc".

• **Always Bcc Myself** With some email accounts, messages sent from your iPhone won't get transferred to the Sent folder on your Mac or PC. If this bothers you, as you'd like to have a complete archive of your mail on your computer, turn on Always Bcc Myself under Settings > Mail, Contacts, Calendars. The downside is that every message you send will pop up in your iPhone inbox a few minutes later. The upside is that you'll get a copy of your sent messages next time you check your mail on your Mac and PC. You can copy these into your Sent folder manually, or set up a rule or filter to do it automatically.

• **Default Account** If you have more than one email account set up on the iPhone, you can choose one to be the default account. This will be used whenever you create messages from other applications – such as when you email a picture from within Photos (see p.140).

• **Organize by threads** With this option enabled, Mail groups all the email exchanges in a given conversation together, with the number on the right-hand side displaying how many messages are in the thread.

• **Signature** Even if you have a sign-off signature (name, contact details, etc) set up at home, it won't show up automatically when you use the same account from the iPhone. To set up a mail signature for your iPhone, tap Settings > Mail, Contacts, Calendars > Signature and then enter your signature.

   In some cases, you'll find extra options under the specific accounts listed at the top of Settings > Mail, Contacts, Calendars. For example:

• **Mail days to sync** Microsoft Exchange accounts offer this option for increasing or decreasing the number of days' worth of emails displayed on the iPhone at any one time.

• **Archive messages** Available to Gmail and iCloud users, this option lets you choose which behaviour you'd prefer when you swipe over a message or tap the middle icon at the bottom of the screen while reading: either to delete the message (i.e. send it to the Trash) or archive it (i.e. send it to your Archive or All Mail folder). If set to delete rather than archive, you can still move messages to your Archive by tapping 🗂 and choosing your Archive or All Mail folder.

Email problems

You can receive but not send

If you're using an email account from an Internet Service Provider, you might find that you can receive emails on the iPhone but not send them. If you entered the details manually on the iPhone, go back and check that you inputted the outgoing mail server details correctly, and that your login details are right.

If that doesn't work, contact your ISP and ask them if they have an outgoing server address that can be accessed from anywhere, or if they can recommend a "port" for mobile access. If they can, add this number, after a colon, onto the name of your outgoing mail server – which you'll find at Settings > Mail, Contacts, Calendars > [the relevant account] > Account Info > SMTP. For example, if your server is smtp.myisp.com and the port number is 138, enter smtp.att.yahoo.com:138

If you have more than one email account set up on your iPhone, you could also try using an alternate outgoing mail server. To make sure that they are all available to be used by any given account, select that account, as above, tap SMTP and then toggle the other servers on within the Other SMTP Servers area of the panel.

> **→ TIP** If you're ever having problems with the Mail app, remember that in most cases you can also log in to your email directly via the web using Safari.

There are messages missing

The most likely answer is that you have a POP-based email account and you downloaded them to your Mac or PC before your iPhone had a chance to do so. Most email programs are set up to delete messages from POP email servers once they've successfully downloaded them. However, it's easy to change this. First, open your mail program on your Mac or PC and view the account settings within Preferences or Tools. Click the account and look for a "delete from server" option, which is usually buried somewhere under Advanced. Tell the program to delete the messages one week after downloading them. This way your phone will have time to download each message before they get deleted. It also means you'll have access to more of your messages when checking your mail via the web.

12

Calendars

How to sync and use calendars on the iPhone

More and more of us are switching from paper to digital diaries – not just for work, but for life, too. The iPhone is helping to accelerate this trend by making it easy not just to schedule and manage appointments on the go – but also to view, edit and share multiple calendars and to synchronize them with laptops, iPads and other devices.

Setting up and syncing calendars

Although it's possible to use the iPhone as a standalone diary, most users will want to sync their iPhone with their existing calendars on their Mac, PC, iPad or Google account. Here's how it's done.

• **iCal and iCloud** If you've already got calendars on your Mac in iCal – or on your iPad – the simplest way to sync is to activate the iCloud calendar option on both your iPhone and your Mac or iPad. See p.40.

• **Outlook and iTunes** If you use iCal but prefer not to use iCloud for some reason, or you have your calendars in Outlook on a PC, it's still possible to sync them via iTunes (see p.42). Simply connect the phone to the computer (either via Wi-fi or USB), open iTunes, click the iPhone's icon and switch on Calendar Syncing under the Info tab.

• **Google Sync** If you use Google Calendars, this can be easily set up on your iPhone by setting up your Gmail account and switching on calendars (see p.104). The downside is that – at the

time of writing at least – you'll only be able to see the mail calendar linked to your account, not any others that you've created, shared or subscribed to. If you want to get around this, one option is to use Google Sync, which bypasses the problem by serving up Google email, contacts and calendars via a Microsoft Exchange system. To get started, follow the (slightly fiddly) instructions at:

Google Sync google.com/sync

An alternative is to use a third-party app such as CalenGoo (see p.130) or to use the – actually rather good – Web app version accessible with Safari at the usual Google calendar address.

> **→ TIP** If you're struggling to set up a Gmail calendar that you only need to be able to view, rather than edit, consider subscribing to it instead – see p.128.

• **Corporate account** Most corporate calendars are linked with a Microsoft Exchange email account. To import and sync these calendars, simply set up the email account and switch on calendars as described on p.119.

Once syncing with iCloud or iTunes is working, calendar data is merged between your computer and phone, so deletions, additions or changes made in one place will immediately be reflected elsewhere (or the next time you connect to iTunes if you're doing it that way).

Subscribing to calendars

The options described above are all about setting up calendar accounts in ways that enable you to add and edit events, as well as view events that already exist. However, it's also possible to "subscribe" to a calendar created by someone else. This is particularly handy for things such as sports fixtures and public holidays. A quick Google search will reveal lots of options. When you've found one you want, grab the subscription address – which should be clearly shown on the website highlighting the calendar – and then add it to your iPhone as a subscription. This is done by tapping Settings > Email, Contacts, Calendar > Add Account > Other.

Calendars on the iPhone

The iPhone Calendars app is very simple to use and doesn't require much explanation. You can view your schedule by month or day, or in a list. There's no proper week view, though if you rotate the phone you can see today, yesterday and tomorrow and slide left and right to view other days.

To add a new event, tap **+**, enter whatever data you like, and tap Save. To edit or delete an existing event, tap the relevant entry and use the Edit button or Trash icon.

Multiple calendars

You can have as many separate calendars on the iPhone as you want. This is useful if you're syncing with a work and home account, for example, or if you want to have a specific calendar for a particular area of your life – childcare, for example, which might be shared between two parents. Each calendar has its own colour to make them easy to differentiate.

When you create a new event, you can choose which calendar it will be added to. If you don't specify, the event will be added to your default account, which you can set under Settings > Mail, Contacts, Calendar. To change an existing event from one calendar to another, tap it, followed by Edit, and use the Calendar option.

Notifications

You can set an alert to remind you of an impending event either as it happens or a certain number of minutes, hours or days beforehand. To set up how these alerts appear – for example, in the middle or the top of the screen – tap Settings > Notifications > Calendar.

> **→ TIP** To quickly view recent calendar notifications without opening the Calendar app, swipe down from the top of the iPhone screen to reveal the notification area.

Invites

Most types of calendar accounts support invitations. The person who created an event can invite other people to whom the event is relevant; they accept  and the event appears in their calendar too. To invite someone to an event, tap it followed by Edit > Invitees and add some email addresses.

Incoming invites can be accepted or declined via the invitations tray at the bottom-right of the Calendar app. The number of unanswered invitations shows up as a badge on the app's Home Screen icon.

→ TIP You can decide whether or not incoming calendar invites flash up on your screen under Settings > Mail, Contacts, Calendar > Invite Alerts.

Calendar apps

Just as it's possible to subscribe to calendars for public holidays, sports fixtures and so on, it's also possible to download calendar apps that provide further information and interactivity, not just dates. There are hundreds of examples – from apps created by specific sports teams and TV shows to apps for keeping track of lunar, solar and even fertility cycles. So if you're interested in anything time-specific, search the app store to see what's on offer.

In addition, you may also want to check out the various apps designed to substitute for Apple's own Calendar app. Two good options include:

CalenGoo

This popular app (pictured) is designed to offer a better experience for Google calendars. It syncs all your calendars and offers a wide range of views.

Fast Calendar & Tasks

This app syncs in the same way as the regular Calendar app but offers a different interface designed for flexibility and speed. Among other things, you can see events spread over multiple days.

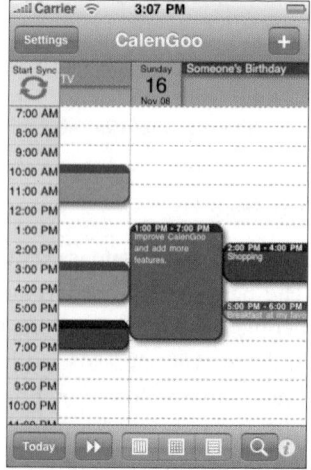

Camera

13

Camera & photos

Pictures in your pocket

The iPhone serves both as a stills and video camera in itself and as a photo album to show off your digital photo collection – including pics taken with other cameras. This chapter offers some tips on using the iPhone's built-in camera for shooting video and stills before explaining how to get your existing images onto your phone.

The iPhone camera

All iPhones have a built-in digital camera, and with each new model the spec and capabilities of this tool have increased. The iPhone 4 has a 5-megapixel stills camera on the back, while the iPhone 4S's camera offers an impressive 8 megapixels. Both models can also shoot HD video. In addition, they each have a second front-facing lens that can be used for stills, video and Apple's much-hyped FaceTime calls (see p.91). Though the front-facing cameras do not offer the same image quality as the ones on the back, they are very useful… and fun.

Shooting stills

Launch the Camera app by tapping its icon on the Home Screen.

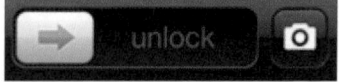

> ➜ **TIP** You can also very quickly access the Camera app straight from the iPhone's Lock Screen by double-tapping the Home button and then tapping to the right of the "unlock" slider.

To take a shot, aim and hold down the "shutter" button ; release your finger at the moment you want to take the shot. As an alternative to using the onscreen "shutter" button, you can also use the physical "volume-up" button on your iPhone to take pictures. This feels particularly natural in landscape mode.

Depending on which model you have, try some, or all, of the following:

• **Hold it steady** The iPhone takes much better pictures when it's held steady, and when the subject of the picture is not moving. Try leaning on a wall, or putting both elbows on a table with the iPhone in both hands, to limit wobble.

• **Keep the lens clean** Given that phones spend a large amount of their lives in pockets and bags, their camera lenses can get pretty smeary. If you find that many of your snaps look a little cloudy, invest in a microfibre cloth and give the lens a quick wipe before you point and shoot.

• **Stay focused** The iPhone 4 and iPhone 4S can both auto-focus. To choose exactly which object or person you want the camera to focus on, tap on the screen to position the square focus target.

• **Auto Exposure lock** If you tap and hold, the square target starts to flash to indicate that the exposure for your next snap is locked. Tap anywhere else on the screen to unlock the exposure and focus the camera as normal.

• **Shoot outside** The iPhone takes much better images outside, in daylight, than it does inside or at night. However, too much direct light and the contrast levels hit the extremes.

> → **TIP** Don't forget that you can rotate the iPhone to shoot both still shots and video in landscape mode too.

• **Light source** Make sure that the light source is behind the camera and not behind the subject of the photograph.

• **Use the flash** If you have an iPhone with a flash, tap the onscreen ⚡ icon, to access the device's built-in flash controls. If you do take shots inside, expect to get better results in light that tends towards white, as opposed to that produced by more yellowy bulbs.

Flash is Disabled
The iPhone needs to cool down before you can use the flash.

OK

> → **TIP** If you ever get an error message from your iPhone's flash, tap the Home button, then relaunch the Camera app and everything should be fine.

• **Align with the grid** Tapping Options at the top of the screen reveals the Grid controls – useful for composing and aligning your subjects.

> → **TIP** After taking a shot, swipe to the right to quickly preview the photo.

• **Using HDR** The Options button also gives you access to High Dynamic Range mode. With this mode enabled, each tap of the shutter generates two images: one normal photo and one HDR photo that, in theory at least, has been auto-processed to create a range of colour and light intensity that more realistically resembles the way we see the world.

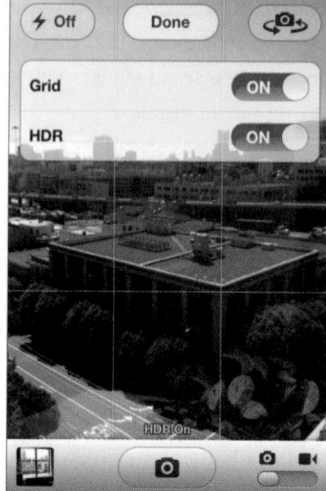

• **Get close and zoom** To take a decent portrait, you'll need to be within a couple of feet of the subject's face. This way more pixels are devoted to face rather than background; in addition, the exposure settings are more likely to be correct. Reverse-pinch the screen to zoom your frame, and to reveal the zoom slider.

> **→ TIP** If you want to get really close to your subject, consider combining your iPhone's camera with a magnifying lens or microscope. You can then make some new friends in this Flickr group: flickr.com/groups/424440@N23

• **Self portraits** Tap the "camera switch" button, top-right, to use the front-facing camera to take pictures of yourself. Obviously you will lose the option of the flash, as it is facing the other way.

> **→ TIP** One extra thing you can do with the camera is to shoot pictures of friends and family and then assign them to relevant entries in the Contacts lists via the option behind the ✉ button. Annoyingly, though, you have to create a contact entry first (see p.108); you can't create a new contact directly from a picture.

Shooting video

To switch between stills and video shooting, toggle the little switch, bottom-right. Though you can't enable the zoom feature for video, from here on the process is pretty much the same as for stills, but with the addition of a glowing red light on the stop/start button to let you know when you are shooting.

As with stills, you can also use the physical volume-up button to stop and start recording. And, as you might expect, you can tap the "camera switch" button, top-right, to use the front-facing camera to film yourself – great for recording video blog posts.

> **➜ TIP** When shooting video in low light, switch the ⚡ button to On for continuous illumination.

The Camera Roll

The images and videos you take are saved together in the so-called Camera Roll, which can be found, when using the camera, by tapping the preview of your last shot, bottom-left. Alternatively, look within the Photos app. Videos appear here with a ⊙ icon in their centres. Tapping the screen reveals additional controls and a scrubber bar for moving back and forth through the footage.

Note the controls at the top for toggling between photos and videos, and the ▶ button at the bottom, which is used to kick-start a slideshow of all the photos in the Camera Roll. To delete, share, copy or print (see p.140) photos from here, tap the ✎ button and then make selections by tapping the previews on the grid.

Putting existing pics on an iPhone

The iPhone can be loaded up with images from your computer. iTunes moves them across, in the process creating copies that are optimized for the phone's screen, thereby minimizing the disk space they occupy. iTunes can move images from an individual folder anywhere on your computer, or from one of three supported photo-management tools:

• **iPhoto (Mac)** apple.com/iphoto
Part of the iLife package, which is free with all new Macs (and available separately for $14.99/£10.49 from the Mac App Store). Version 4.0.3 or later will sync photos and videos with an iPhone, but version 6 is better, enabling you to view pictures according to the faces in them.

> **➜ TIP** If you sync your iPhone with iPhoto on a Mac, when you connect the phone to your Mac, iPhoto may automatically launch and offer to import recent snaps taken with your camera phone. (If your Camera Roll is empty, this won't happen.) The way around this is to open iPhoto > Preferences and from the General tab, choose "Connecting camera opens: no application" from the dropdown list of options.

• **Aperture (Mac)** apple.com/aperture
Apple's professional photo suite can do the job, but is pricier ($79.99/£54.99 from the Mac App Store) and only recommended to the dedicated photographer.

• **Photoshop Elements (PC)** adobe.com/photoshopelements
This Adobe app is similar to iPhoto, but with far more editing tools and a $79.99/£64.81 price tag. You'll need version 3.0 or later.

All these applications offer editing tools for colour balance and so on, and allow you to arrange your images into "albums", which will show up in a list on your iPhone. If you prefer, though, you can keep your photos

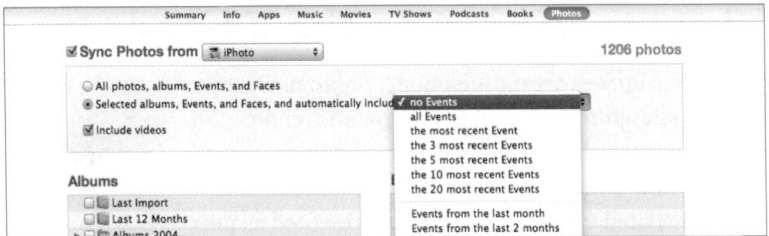

within a standard folder, such as the My Pictures folder in Windows, or the Pictures folder in OS X. Any subfolders will be treated as albums.

To get started, connect your iPhone and look for the Photos tab in iTunes. Check the "Sync Photos" box, then choose your application or folder. If you sync with a specific folder, any subfolders will show up on your iPhone as albums within the Photos app. Note the dropdown menu option that lets you choose exactly how far back you sync – useful if you have a sizeable photo collection.

iCloud and Photo Stream

With Apple's iCloud service (see p.40) you can synchronize photos (but not videos) across various Apple devices and computers. It doesn't sync all the photos in your collection to your iPhone (there probably wouldn't be room), only the most recent 1000 shots that have been snapped on your Apple devices or added to the albums on your computer. It operates via the iCloud servers, but only over Wi-Fi or a wired network, so don't expect it to work over your carrier's data network. As a bonus, the 1000 photos don't count against your allotted iCloud storage limit.

To get started, make sure iCloud Services are enabled on your iPhone (Settings > iCloud), turn on Photo Stream in Settings > Photos, then look for the Photo Stream album within the Photos app.

> → **TIP** iCloud-synced images are automatically deleted from the Photo Stream after thirty days, so if you want to make sure specific snaps stay on your iPhone longer, view them in the Photo Stream album and tap 🖼 > Save to Camera Roll.

Viewing images on the iPhone

Once your images are on the phone, photo navigation is very straightforward. Tap Photos and choose an album – or tap Photo Library to see the images in all albums. You can also choose Places to see photos that have associated GPS location data represented as pins on a map.

> **→ TIP** If you have synced photos from iPhoto or Aperture based on the Faces feature recognition tool, tap Faces to view the list. Unfortunately, the Photos app only displays synced Faces sets and cannot add to them as you take new photos.

When viewing a set of images:

• **Share** Tap ☑, highlight the images you want to send and then tap Share to add them to an email or MMS message.

• **Print** Tap ☑, highlight the images you want to print, then tap Share > Print to send them to a local AirPrint printer (see p.246).

• **Copy** Tap ☑, highlight the images you want and then tap Copy. The images can then be pasted into an email, text message or other document by tapping at the insertion point and choosing Paste from the pop-up panel.

> **→ TIP** To create a new album within the Photos app, tap ☑, make selections, and then tap Add to... > Add to New Album.

When viewing individual images:

• **"Flick" left and right** to move to the previous or next photo.

• **Zoom in and out** Double-tap or "stretch" and "pinch" with two fingers.

• **Rotate the iPhone** to see the picture in landscape mode.

• **Hide or reveal the controls** Tap once anywhere on the image.

• **More options** Tap ✉ to assign an image to a contact, send as an MMS or email, print, Tweet, or use it as wallpaper.

• **Start a slideshow** Tap ▶, and then choose a Star Wars-esque transition and music if you want it. By default, the phone will show each photo for three seconds, but you can change this by heading to Settings > Photos. The same screen lets you turn on Shuffle (random order) and Repeat (so that the slideshow plays around and around until you beg it to stop). You can also connect to a TV via a cable (see p.185) to play your slideshows on a big screen.

• **AirPlay** If you see the 🔲 icon at the bottom of the screen, then it also means you have the option of viewing your photos via an Apple TV device on the same network.

Native photo editing

When viewing an individual image in Photos tap the Edit button to see the available options. From left to right:

• **Rotate** Each tap on the arrow icon spins the image a further 90 degrees anti-clockwise. Tap Save when you are done.

• **Auto-Enhance** This automatic setting (denoted by a magic wand icon) does a pretty good job of tweaking sharpness, contrast, brightness and levels to get the most out of each shot. Tap Save if you are happy with the results.

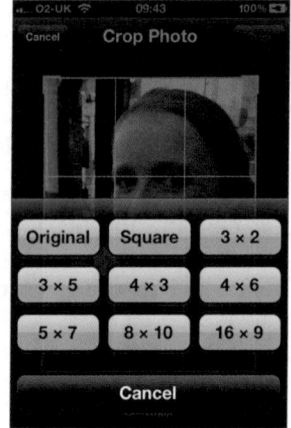

• **Red-Eye** Tap the Red-Eye button, then tap each eye to remove that devilish glow. Tap the eyes again to undo. When you're done, tap Apply, and then Save.

• **Crop** Drag the corners of the onscreen grid to reframe manually, and use two fingers to twist the image if it needs a little realignment. Alternatively, tap the Constrain button to reveal nine presets. When you're ready, tap Crop.

Native video editing

You can snip the ends off video clips straight from a video preview in either Photos or the Camera Roll. Simply drag in the two end points on the scrubber bar (you'll see it turn yellow when you are in trim mode) and then tap the Trim button, top-right, when you're done.

➜ **TIP** For more sophisticated third-party photo and video editing tools available in the App Store, see the next chapter.

14

Apps: Camera

As we've already seen, there is a lot you can do with the iPhone's camera straight out of the box, but there's also plenty more to be discovered. Whether you want to craft near-professional quality photos, or re-create the look and feel of a bygone age of film-making, there are hundreds of apps out there to help you get things done. Here are a few of our favourites.

Camera apps

CameraBag
A really nice collection of filters and visual effects to enhance your iPhone photography. It reproduces everything from infrared to fisheye to "magazine" style.

Camera+
This is an excellent, fully featured alternative to the iPhone's built-in Camera app. Its special image stabilizer control works really well to stop the wobble.

FatBooth

This little app shows what you would look like if you were more interested in pies than iPhones. The same developers also make AgingBooth.

Hipstamatic

Relive those analogue days of oversaturated, grainy snaps with this very popular app. It allows you to share your creations via Facebook and Twitter and has some really nice tools for organizing your images into "stacks".

AutoStitch

An excellent app for creating stitched panoramas from multiple shots. It then allows you to crop your creations and save them to the iPhone's Camera Roll.

Adobe Photoshop Express

This is Adobe's iPhone-friendly version of their popular Photoshop desktop application. It's really easy to use and has some nice effects, as well as a useful crop tool that offers a supercharged version of the native iPhone tool.

Tiltshift Generator

This app has an interesting toolkit of filters and blurs that re-create some of the hallmark effects of 1960s toy cameras, which used cheap lenses and were largely made from plastic.

You can use this app to create interesting focus effects within existing images.

Tom Ang

From professional photographer and writer Tom Ang, this is one of the best photographic guidance apps available for the iPhone. As well as a wealth of general tips for framing and composing your shots, there is a great filtering tool for choosing the information you need based on the subject matter and conditions ... all handily available on your iPhone.

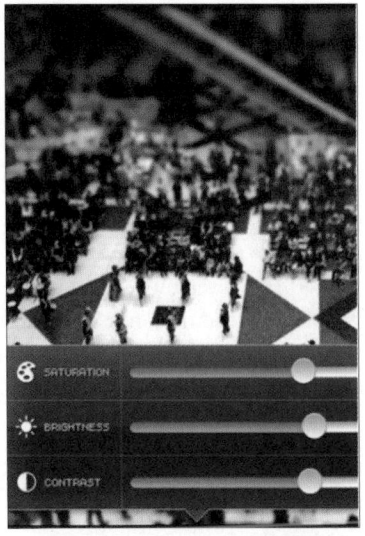

Photogene

The built-in editing tools of the Photos app don't come close to matching what's on offer here. You can dive in and adjust curves and colour graphs, add art effects and even add speech bubbles and captions.

Snapseed

Another top-class app for adjusting photos and images to create some very professional results.

Video apps

Video Camera (for iPhone 2G and 3G)

That's right, a video camera app for pre-3GS iPhones that don't support video-recording straight out of the box.

iMovie

Great for adding sophisticated editing effects, transitions, themes, music and the like. Though some of the themes might seem a little cheesy, the timeline and transition tools work well on the small screen.

iSuper8

A really nice little app for shooting Super 8-style footage. Unlike the original format, you get the bonus of audio thrown in.

Online image posting from the iPhone

Aside from Apple's own Photo Stream tool (see p.139), there are hundreds of apps for photo sharing and uploading waiting to be discovered in the App Store. Flickr, the world's most popular photo-sharing site, has an excellent free app that account-holders can use on their iPhones. Pixelpipe, meanwhile, allows you to upload pictures to Flickr, Picasa, Facebook, YouTube and dozens more sites, all from one place. Instagram is another decent choice; it features a wealth of filters and effects for styling your images as well.

It's also worth noting that the Facebook app (see p.221) and nearly all Twitter-interfacing apps also allow you to post images using your camera. If you have entered the appropriate account details within Settings > Twitter on your iPhone you can also post to Twitter directly from both the Camera Roll and albums within Photos. Simply tap the ✉ button and choose Tweet.

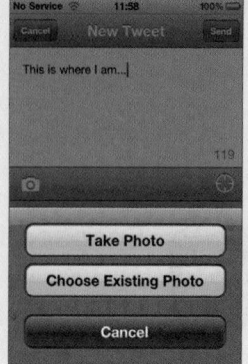

Music & video

iTunes prep

Preparing music & video files to sync with the iPhone

Downloading music and video from the iTunes Store (either direct to the iPhone or to your computer) is all well and good, but if you already own the CD or DVD, there's no point in paying for the same content again. "Ripping" CDs and DVDs to get them into iTunes (and in turn onto the iPhone) is easy, but it's worth reading through this chapter even if you've done it hundreds of times, as various preferences and features are easy to miss.

Importing CDs

To get started, insert any audio CD into your Mac or PC. In most cases, within a few seconds you'll find that the artist, track and album names – and maybe more info besides – automatically appear. If your song info fails to materialize (and all you get is "Track 1", "Track 2", etc), or you want to edit what has appeared, either click into individual fields and type, or select multiple tracks and choose File > Get Info to make your changes.

Settings and importing

Before importing all your music, have a look through the various importing options, which you'll find behind the Import Settings button on the General pane of iTunes Preferences. These are worth considering early on, as they relate to sound quality and compatibility. The iPhone can play MP3 and AAC files (up to 320 kbps) as well as Apple Lossless, AIFF, Audible and WAV files (see box opposite).

The bitrate is the amount of data that each second of sound is reduced to. The higher the bitrate, the higher the sound quality, but also the more space the file takes up. The relationship between file size and bitrate is basically proportional, but the same isn't true of sound quality, so a 128-kbps track takes half as much space as a 256-kbps version, but the sound will be only marginally different.

> **→ TIP** To rip multiple tracks as one, simply select them and click Advanced > Join CD Tracks before you import.

Take a quick look at the other options on offer, but don't worry too much as the defaults will do just fine. That said, the iPhone only has a mono speaker (but does have a stereo headphone jack), so if capacity is an issue you could rip mono versions of your songs to use on it – they'll take up half the space of stereo versions.

When you are ready to import, hit the Import button in the bottom-right corner of the iTunes window.

Import Settings
Import Using:	MP3 Encoder ⬍
Setting:	Higher Quality (192 kbps) ⬍
Details
96 kbps (mono)/192 kbps (stereo), normal stereo, optimized for MMX/SSE, using MP.

A Rough Guide to music file formats

Music can be saved in various different file formats, just like images (bitmap, jpeg, gif, etc) and text documents (doc, txt, rtf, etc) can be. When you import a CD to iTunes, you can pick from AAC, MP3, Apple Lossless, Wav and Aiff. Here's the lowdown on each:

MP3 [Moving Pictures Experts Group-1/2 Audio Layer 3]

Pros: Compatible with all MP3 players and computer systems. Also allows you to burn high-capacity CDs for playback on MP3-capable CD players.

Cons: Not quite as good as AAC in terms of sound quality per megabyte.

File name ends: .mp3

AAC [Advanced Audio Coding]

Pros: Excellent sound quality for the disk space it takes up.

Cons: Not compatible with much non-Apple hardware or software.

File name ends: .m4a (or .m4p for protected files from the iTunes Store)

Apple Lossless Encoder

Pros: Full CD sound quality in half the disk space of an uncompressed track.

Cons: Files are very large and only play on iTunes, iPhones, iPads and iPods.

File name ends: .ale

AIFF/Wav [Audio Interchange File Format]

Pros: Full CD sound quality. Plays back on any system.

Cons: Huge files.

File name ends: .aiff/.wav

Converting one music file format to another

iTunes allows you to create copies of imported tracks in different file formats. This is great for reducing the size of bulky WAV, AIFF or Apple Lossless files, or for creating MP3 versions of songs that you want to give to friends who have non-Apple music players or phones. Be warned, however, that re-encoding one compressed file format (such as AAC) into another (such as MP3) will damage the sound quality somewhat.

To create a copy of a track in a different format, first specify your desired format and bitrate on the iTunes Import Settings, within the General pane of iTunes Preferences. Then, close Preferences, select the file or files in question in the main iTunes window and choose Create MP3/WAV/AAC Version from the Advanced menu.

When copying high bitrate songs to your iPhone, you can set iTunes to convert them automatically to 128 kbps by checking the appropriate box under the Info tab of your iPhone's options panel in iTunes.

Importing DVDs

Just as with music, before you can transfer video files to your iPhone, you first have to get them into iTunes. In most cases, it's perfectly possible to do this from DVD, though in some countries this may not be strictly legal when it comes to copyrighted movies. As long as you're only importing your own DVDs for your own use, no one is likely to mind. The main problem is that it's a bit of a hassle. A DVD contains so much data that it can take more than an hour to "rip" each movie to your computer in a format that'll work with iTunes and an iPhone. And if the disc contains copy protection, then it's even more of a headache.

> ➡ **TIP** Ripping Blu-ray discs is a prohibitively complex and time-consuming process (possible only with a PC). Better to hunt down DVDs that come with an accompanying digital version.

Using HandBrake

Of the various free tools available for getting DVDs into iTunes, probably the best is HandBrake, which is available for both Mac and PC. Here's how the process works:

• **Download and install HandBrake** from handbrake.fr

• **Insert the DVD** and, if it starts to play automatically, quit your DVD player program.

DVD copy protection

DVDs are often encrypted, or copy protected, to stop people making copies or ripping the discs to their computers. PC owners can use a program such as AnyDVD (slysoft.com) to get around the protection, while Mac owners trying to get encrypted DVDs into iTunes will need to grab a program such as Fast DVD Copy (fastdvdcopy.com). This allows you to make a non-protected copy of the movie, which you can then turn into an iPad-formatted version using HandBrake. Note that, in some countries, it may not be legal to copy an encrypted DVD.

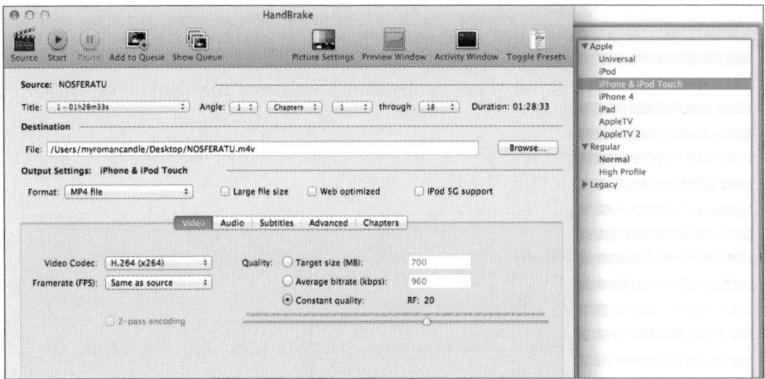

• **Launch HandBrake** and it should detect the DVD (it may call it something unfriendly like "/dev/rdisk1"). Press Open, and wait until the application has scanned the DVD.

• **Choose iPhone-friendly settings** Choose the iPhone option that relates to your model from the Presets menu.

• **Check the source** It's also worth taking a quick look at the Title dropdown menu within the Source section of HandBrake. Choose the one that represents the largest amount of time (say 01h22m46s) as this should be the main feature. If nothing of an appropriate length appears, then your DVD is copy protected.

• **Subtitles** If it's a foreign-language film, set Dialogue and Subtitles options from the dropdown menus behind the Audio & Subtitles tab.

• **Rip** Hit the Start button at the bottom of the window and the encoding will begin. Don't hold your breath.

➔ **TIP** Some DVDs feature promotional codes that entitle you to a free iTunes digital version of the same movie.

Supported video formats

To get technical for a moment, note that the various video formats that the iPhone supports include:

H.264-encoded video up to 720p at 30 frames per second, with AAC-LC audio up to 160 kbps, 48 kHz, in m4v, mp4 and mov formats.

MPEG-4 up to 640x480 pixels, 2.5 Mbps, with AAC-LC audio up to 160 kbps, 48 kHz.

Motion JPEG (M-JPEG) Avi-wrapped and up to 1280x720 pixels, 35 Mbps, 30 frames per second with ulaw PCM stereo audio. (This is the video format often used by digital cameras that can also shoot video.)

If you find that you have files on your computer that don't meet these criteria, either convert them within iTunes or use HandBrake, as previously described.

• **Drop the file into iTunes** Unless you choose to save it somewhere else, the file will eventually appear on the Desktop. Drag the file into the main iTunes window. This should create a copy of the new file in your iTunes Library allowing you to then delete the original file from your Desktop.

> ➜ **TIP** To learn about importing from other music and video sources, see *The Rough Guides to iPods & iTunes*.

Converting video files in iTunes

If you find yourself with music or video files in iTunes that can't be copied across to the iPhone when you sync (perhaps they are the wrong file type or, in the case of music files, have too high a bitrate), you can easily convert them to an iPhone-friendly format.

To do this for a video file, select it within iTunes and choose Advanced > Create iPod or iPhone Version. In the case of an audio file, the option is determined by the Import Settings that you define within iTunes Preferences. So, for example, if iTunes is set to import files as MP3, you will be offered the option to Create MP3 Version within the Advanced menu.

Recording from vinyl or cassette

If you have the time and inclination, it's perfectly possible to import music from analogue sound sources such as vinyl or cassette into iTunes and onto your iPhone. For vinyl, you could buy a USB turntable (such as those from Ion or Kam), but this isn't strictly necessary. With the right cables, you can connect your hi-fi, Walkman, minidisc player or any other source to your computer and do it manually.

• **Hooking up** First of all, you'll need to make the right connection. With any luck, your computer will have a line-in or mic port, probably in the form of a minijack socket (if it doesn't, you can add one with the right USB device; ask in any computer store). On the hi-fi, a headphone socket will suffice, but you'll get a much better "level" from a dedicated line-out.

• **Choose some software** Recording from an analogue source requires an audio editing application. You may already have something suitable on your computer, but there are also scores of excellent programs available to download. Our recommendations are GarageBand for Mac users, which anyone with an Apple machine purchased in the last few years will already have, and Audacity, which is available for both PC and Mac; it's easy to use and totally free:

Audacity audacity.sourceforge.net
GarageBand apple.com/garageband

> ➜ **TIP** Roxio produce an excellent "Easy LP to MP3" kit that includes software and all the necessary cabling you need. Find out more from www.roxio.co.uk.

• **Recording** Connect your computer and hi-fi as described above, and switch your hi-fi's amplifier to "Phono", "Tape" or whichever channel you're recording from. Launch your audio recorder and open a new file. The details from here on vary according to which program you're running and the analogue source you are recording from, but, roughly speaking, the procedure is the same.

You'll be asked to specify a few parameters for the recording. The defaults (usually 44.1 kHz, 16-bit stereo) should be fine. Play the loudest section of the record to get an idea of the level. A visual meter should display the sound coming in.

If your level is too low, tweak your line-in volume level: on a Mac, look under System Preferences > Sound; on a PC, look in Control Panel.

When you're ready, press "Record" and start your vinyl, cassette or other source playing. When the song or album is finished, press "Stop". Use the "cut" tool to tidy up any extraneous noise or blank space from the beginning and end of the file; fade in and out to hide the "cuts", and, if you like, experiment with any hiss and filters on offer.

• **Drop it into iTunes** When you are happy with what you've got, save the file in WAV or AIFF format, import it into iTunes (choose Import… from the File menu), convert it to AAC or MP3 (see p.151) and delete the bulky original from both your iTunes folder and its original location.

Managing files in iTunes

Once you start digging around within the Music and Movies sync tabs in iTunes, it will soon become clear that the easiest way to manage your content for syncing to the iPhone is by using playlists:

• **Regular playlists** To create a playlist, hit the New Playlist button (the **+**) at the bottom-left of the iTunes window. Then drag individual songs into the new list or add entire albums, artists or genres in one fell swoop. You can also create a new playlist by dragging selections into the sidebar, or by selecting a bunch of material and choosing File > New Playlist.

• **Smart Playlists** Rather than being compiled manually by you, these are put together automatically in accordance with a set of rules, or conditions, that you define. It might be songs with a certain word in their title, or a set of genres, or the tracks you've listened to the most – or a combination of any of these kinds of things. What's clever about Smart Playlists is that their contents will automatically change over time (assuming you tick the "Live" updating box), as relevant tracks are added to your Library or existing tracks meet the criteria by being, say, rated highly. To create a new Smart Playlist, look in the File menu or click the New Playlist button while holding down Alt (on a Mac) or Shift (on a PC) – you'll see the **+** button change into a cog.

Editing track info

A common niggle on the iPhone and in iTunes is that you might end up with inconsistent labelling of track information. For example, you might find that you have one album by "Miles Davis" and another by "Davis, Miles". Thankfully, you can quickly edit this information. Simply select one or more tracks – or even whole albums, artists, composers or genres – and choose Get Info from the file menu.

> **➜ TIP** There are also applications that you can download to help with the clean-up process and find missing artwork. TuneUp (tuneupmedia.com) is a great choice for both Windows and Mac machines.

Home Sharing

Home Sharing is an iTunes feature that lets you stream the contents of your library across a home network to your other computers, iPads, iPhones and the iPod Touch. To get things started, open iTunes (you'll need version 10.2 or later), then Preferences and, under Sharing, check the "Share my library on my local network" box. Next, click Enable Home Sharing in the iTunes Advanced menu. You'll then be prompted to enter your Apple ID – this will be the same one that you use in the iTunes and App Stores.

You are now ready to enable Home Sharing on your iPhone (look within Settings > Music, and also in Settings > Video).

Syncing with the iPhone

Once all your audio and video files are organized and ready to sync, connect your iPhone to your computer using a cable and check the boxes for the content you want to move across within the Music, Movies and TV Shows tabs of iTunes.

Take note of the options at the top of the Music tab. If you expect to use the iPhone's Memos app (see p.239), check the box to include voice

iTunes Wi-Fi Sync

While your iPhone is connected to iTunes via a cable, take the opportunity to enable Wi-Fi syncing so that in future you can sync without having to make the physical connection. In iTunes, highlight your iPhone in the sidebar and then under the General tab check the "Sync with this iPhone over Wi-Fi" box, hit Apply and then hit Sync.

Next, disconnect your iPhone, make sure it is wirelessly connected to the same Wi-Fi network as the computer, and navigate to Settings > General > iTunes Wi-Fi Sync. Tap Sync Now to check that it is working.

You can now either kick-start a sync from your phone (as above) or go into iTunes on your computer, make changes to the sync parameters and kick-start things from there... again, assuming that both phone and computer are on the same Wi-Fi network.

memos in your sync. Don't check the "Automatically fill free space…" box, as this will limit your ability to download additional apps and music when out and about.

If you try to load more music onto the iPhone than there is space for, iTunes will ask if you want to create a playlist of the appropriate size and set it to sync. If you answer yes, iTunes will randomly fill a new playlist which you can add to and remove from in the usual way.

If you prefer, instead of having the iPhone sync with a particular play-list or set of playlists (see p.156), you can move content to your iPhone by simply dragging-and-dropping. To opt for this approach, click the Summary tab and check the box labelled Manually Manage Music and Video; then hit Apply.

Now you can drag tracks, artists, playlists or even whole genres (if you have the space) straight onto your iPhone's icon from within iTunes. Clicking the ▶ icon to the left lets you see the contents of your iPhone in more detail, allowing you to drag music into specific playlists.

With the iPhone set up in this way, removing music is also handled manually, but don't worry: deleting a song from your phone will not affect the original in your computer's iTunes library.

Moving music from your iPhone to a computer

Music purchased on iTunes

If you plug your iPhone into a computer other than the one it's paired with, it will allow you to copy onto the computer any content downloaded from the iTunes Store. This option may pop up automatically; alternatively, click Transfer Purchases from… in the File menu whenever an iPhone is connected.

Of course, the music will only play back if the computer in question is one of the five machines authorized for your iTunes Apple ID (see p.169).

Other music

You can't copy music *not* purchased at the iTunes Store from iPhone to computer. This setup is designed to stop people sharing copyrighted music, but can be a real pain if your computer is stolen or destroyed, and the only version of your music collection you have left is the one stored on your iPhone.

Since the iPod and iPhone became popular, many applications have become available allowing iPhone and iPod-to-computer copying. These have never been formally recognized by Apple, but they've generally worked well enough. Two worth investigating are The Little App Factory's iRip and Kennett Net's Music Rescue:

The Little App Factory thelittleappfactory.com
Kennett Net kennettnet.co.uk

16

The iTunes Store

Buying and renting, direct from the iPhone

The iTunes Store isn't the only option for downloading music and video from the Internet. But if you use the iPhone, it's unquestionably the most convenient, offering instant, legal access to millions of music tracks and music videos, plus a growing selection of TV shows and movies to either buy or rent. Unlike some download stores, the iTunes Store is not a website, so don't expect to reach it with Safari – the only way in is through either iTunes on a Mac or PC, or via the iTunes icon on your iPhone's Home Screen.

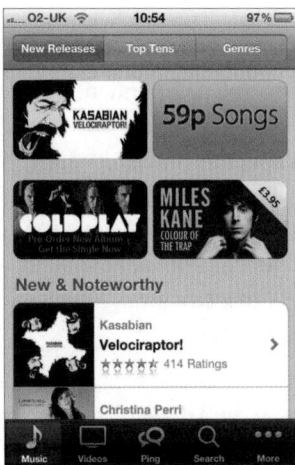

What have they got?

At the time of writing, the iTunes Store boasts more than fourteen million tracks worldwide, plus twenty thousand audiobooks and many thousands of movies, TV shows and podcasts (see box opposite). It claims to have the largest legal download catalogue in the world.

> → **TIP** To access the Store from your computer, click the Store icon in the iTunes sidebar; once downloaded, you can then sync purchased content to the iPhone as and when you need to.

However, there are some glaring music and movie omissions, and it isn't like a regular shop where anything can be ordered if you're prepared to wait a while. As with any download site, everything that's up there is the result of a deal struck with the record label or movie company in question, and several independent record distributors have refused to sign up. So don't expect to find everything you want.

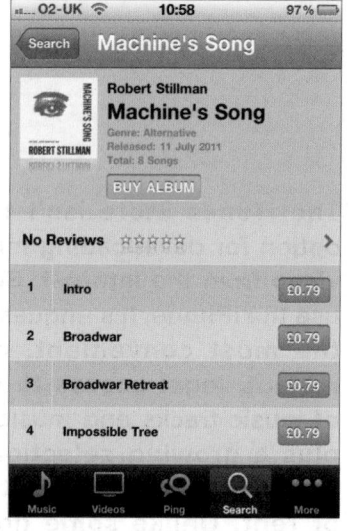

That said, thousands of new music tracks, audiobooks, TV shows and feature films appear week after week, so the situation is getting better all the time.

> → **TIP** The iTunes Store app also gives you access to Apple's Ping social network, which lets you "follow" bands and friends and see what they are listening to. If you want to try it out, open iTunes on your computer, click Store and then Ping.

Podcasts

The best way to understand podcasts is to think of them as audio or video blogs. Like regular blogs, they are generally made up of a series of short episodes, or posts, which are nearly always free. Podcasts often consist of either radio-style spoken content or condensed documentary or chatshow-style video episodes, covering everything from current affairs and poetry to cookery and technology. There are many musical podcasts, too, though there's a grey area surrounding the distribution of copyrighted music in this way.

Podcasts are made available as files (audio or video) that can be downloaded either straight to your iPhone or, alternatively, to a Mac or PC, from which you can sync them across.

Subscribing to podcasts from a Mac or PC

The iTunes Store offers by far the easiest method of subscribing to podcasts. Open iTunes, click iTunes Store in the sidebar and then click the Podcast tab to start browsing for interesting-looking podcasts. When you find one that looks like it's up your street, click Subscribe, and iTunes will automatically download the most recent episode to your iTunes Library. (Depending on the podcast, you may also be offered all the previous episodes to download.)

To change how iTunes handles podcasts, click Podcasts in the sidebar and then the Settings button at the bottom. For example, if disk space is at a premium on your system, tell iTunes to only keep unplayed episodes.

To sync your podcasts over to the iPhone, connect it and look for the options under the Podcasts tab.

If you want to stop your kids accessing podcasts through iTunes on your Mac or PC, check the relevant option under the Parental tab of iTunes Preferences.

Subscribing to podcasts on the iPhone

From the Home Screen, tap iTunes > More > Podcasts and browse just as you would any other department of the store. When you find something you want, tap the FREE button to start an episode downloading.

To play podcasts, by default tap Music > More > Podcasts. For the full story on podcast playback, turn to p.177.

Rolling your own podcasts

Once you've subscribed and listened to a few podcasts you might decide that it's time to get in on the action and turn your own hand to broadcasting. To find out how to get started, visit apple.com/itunes/podcasts/specs.html

iTunes accounts

Though anyone can browse the iTunes Store on the iPhone, listen to samples and watch previews and movie trailers, if you actually want to download anything you need to set up an account and be logged in. If you haven't already done this, it's easily done: either try to buy something, and follow the prompts, or head to Settings > Store and enter the necessary details there. Here, you can also choose whether you want purchases

made on other devices using the same account to be automatically downloaded to your iPhone and, more importantly, whether you want them to be downloaded over the cellular network (best to keep this turned off if you have a capped data plan).

If someone else is already signed in to the Store on the same iPhone, they'll need to sign out first within Settings > Store. Also note that iTunes Store Accounts are country-specific; in other words, you only get to access the store of the country where the credit card associated with the account has a billing address.

> **→ TIP** If you already have an Apple ID, iCloud, iBookstore or AOL login, the same credentials will work here.

Staying secure

By default, the iPhone will remember your account details, making purchasing content almost too easy. This could be asking for trouble (especially if you have kids around). It's safer to enter your password each time you want to buy something, remembering to log out when you're done (scroll to the bottom of any iTunes Store listings page and tap the Account: yourname button).

If that seems like a chore, the alternative is to use a passcode (Settings > General > Passcode Lock) to control who has access to your iPhone and the shopping opportunities therein.

> **→ TIP** If you want to stop your kids accessing the entire Store on the iPhone or being exposed to explicit material, look for the options under Settings > General > Restrictions.

Renting movies

Many of the major movie studios are now making their films (both new releases and back catalogue titles) available to rent via the iTunes Store. Some films are also available in high definition (HD) with a slightly higher rental cost. Once a rented movie file has been downloaded to your iPhone you have 30 days to start watching it, and once you have played even just a few seconds of it, you have a certain period of time to finish it (in the US it's 24 hours, in the UK, 48). When your time runs out, the file miraculously disappears.

A movie rented on the iPhone can't be transferred to a computer or other device to be watched there. You can, however, rent a movie through iTunes on a Mac or PC and then sync it across to the iPhone.

Annoyingly, you need to fully download an entire movie before you can start watching it, which, depending on your Internet connection, could take a few hours. To see how your download is progressing, tap More > Downloads at the bottom of the iTunes Store window.

> **→ TIP** To see trailers, read reviews and find local cinema showing times, download the excellent free Flixster Movies app.

iCloud & iTunes Match

Discussed elsewhere in this book (see p.40), iCloud is Apple's online storage and sync service that, amongst other things, can give you access to your music (but not videos) via Apple's servers, on any of your Apple devices. First launched in the US in late 2011, the music aspect of the service incorporates two slightly different setups, depending on where the tracks in your collection have come from.

iTunes Store purchases

Assuming your iTunes Store credentials are logged within Settings > iCloud on your iPhone, and you have enabled the feature within Settings > Store, then any purchases you make from the iTunes Store on either your computer or any Apple device using the same account, will be automatically synced to all your other devices. This happens by default over Wi-Fi, but can also be enabled over your iPhone's carrier network via Settings > Store (which might not be a great idea if you have a capped data tariff). Store purchases do not count against your allotted iCloud storage capacity.

→ **TIP** To check for any older purchases made using your Store Account that are missing from your iPhone, launch the iTunes app, tap the Purchased tab at the bottom and browse.

Music ripped from CDs or purchased from other stores

This is where iTunes Match comes in. This service gives you cloud access to all the songs (up to 25,000) in your collection that you have *not* purchased from the iTunes Store. Basically, iTunes determines which songs in your collection are already in the cloud (i.e. they are already on Apple's servers because they are sold in the Store) and automatically gives you access to them. Any tracks not "matched" are uploaded from your computer so that they are available for use in your iCloud library.

One of the pluses of this matching process is that your matched iCloud tracks will be at the high iTunes Plus quality (256 Kbps), even if the original tracks you have at home were ripped at a lower bitrate. On the downside, this service will cost you (see apple.com/icloud for the latest rates), and uploading unmatched tracks can take a while. It is also worth noting that while any kind of cloud service is great over Wi-Fi, it can be a real data-hog over a cellular network.

Alternative cloud music services

Spotify

This paid subscription service, and accompanying iPhone app (see p.179), give you access to an all-you-can-eat supply of music over the airwaves. However, you can also sync a selection of your favourite tracks to your iPhone so that you can listen out and about without crippling your network data quota. The desktop application is excellent, and it lets you create playlists and organize your music in much the same way that you do with iTunes; you can even buy the songs outright if you want to. See: spotify.com

Amazon Cloud Drive

Amazon's cloud service gives you 5GB of free storage for all types of file, and any MP3s that you have purchased from their store are stored without eating into your allotted space. There is also the so-called Cloud Player that gives you access to your tracks from any computer connected to the Internet. Though there isn't a Cloud Player app for the iPhone, the web-based version works like a dream, giving you access to your tunes via Safari. Find out more on the website: amazon.com/clouddrive/learnmore

For other apps that give you access to music via your iPhone, turn to p.170.

More in the Store

New features are added to the iTunes Store on a regular basis. Here are a few other things that you might like to explore:

• **Redeeming gift cards** If you are lucky enough to be given one, you can use iTunes Store gift cards and certificates to pay for content in the Store. To redeem a gift card, tap Music, scroll down, tap Redeem and follow the prompts. Your store credit then appears with your account info at the bottom of most iTunes Store screens.

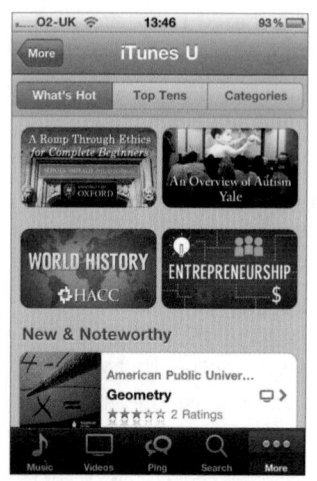

• **iTunes U** iTunes U ("university") makes available lectures, debates and presentations from US colleges as audio and video files. The service is free and has made unlikely stars of some of the more entertaining professors.

• **Freebies** Keep an eye open for free tracks: you get something for nothing, and you might discover an artist you never knew you liked. Also, make the most of the movie trailers on offer in the videos department; simply navigate to a file and tap Preview.

> ➔ **TIP** Purchases sync back to iTunes on your computer when you connect your iPhone; they can be viewed in the Purchased on... playlist that appears in the iTunes sidebar.

• **Genius** Tap the Genius tab to get movie, TV show and music recommendations based upon items you've already bought.

DRM and authorized computers

All tracks in the iTunes Store come without traditional built-in DRM (Digital Rights Management) of the kind that Apple used to use to stop content being copied and passed on. Without the built-in DRM there are no technological barriers to someone distributing the files they have purchased. However, files downloaded from the iTunes Store do contain the purchaser's name and email details embedded as "metadata" within the file. The upshot of this is that files purchased from iTunes and then illegally distributed over the Internet are traceable back to the person who originally shelled out for them. The other thing worth noting about iTunes-purchased files is that they're in the AAC format, so they'll only play in iTunes, on iPods, iPhones, iPads and any non-Apple software and hardware that supports this type of file; unless, of course, you convert them to MP3 first (see p.151).

Apple also use what's known as FairPlay DRM, where content purchased using a specific iTunes Account can only be played on (or synced to, in the case of books and apps) a maximum of five Macs or PCs that are "authorized" for that account. (You can, however, add content to as many iPads, iPods and iPhones as you want; you just won't be able to move it onto a Mac or PC that has not been authorized.)

To manage the machines authorized for your iTunes Account, open iTunes on your Mac or PC and look for the options in the Store menu. To deauthorize the machine you are on (worth doing before you sell or get rid of a computer), choose Deauthorize this Computer. To deauthorize all machines and start afresh (useful if you no longer have access to one or more of your five), choose Store > View My Account > Deauthorize All.

• **Ringtones** The store also has a custom ringtones section, which is discussed on p.48.

• **Ping** Also found behind the More tab within the iTunes app, this is Apple's music-based social networking service. It allows you to create a profile page and share your musical tastes with the world. You can also follow other fans, and artists, which can be a good way to discover new music.

> **→ TIP** To change which departments appear on the lower panel of the iTunes Store app on your iPhone, launch the app and tap More > Edit; then drag and drop the various icons to get the configuration you want.

17
iPhone audio

Music, Remote and other apps

Once your iPhone is loaded up with tunes, you're ready to adjust its on-board audio settings (see box on p.151) and start listening. But there's much more to audio on the iPhone than just the built-in Music app. This chapter will show you a raft of apps that can do everything from turning your iPhone into a remote control to allowing you to tune into your favourite radio station.

The iPhone's Music app

The iPhone's built-in Music app (previously called the iPod app) is easy to use, and is your one-stop shop for playing music, podcasts, audiobooks and music videos. Tap the Music icon on the Home Screen to start.

From there on in, it really is pretty self-explanatory: tap to see listings and then tap a track to hear it. You can also tap [Now Playing] to enter the full-screen "Now Playing" mode.

Editing the options

The More button reveals further options for browsing. Depending on your listening habits, these might be more useful than the default options. For instance, classical music buffs will want instant access to Composers, while radio lovers will want to front-load Podcasts. To replace an existing browse icon with a different one, click More, then Edit, and then simply drag the new icon onto the old one. You can also drag the icons at the bottom into any order.

Now Playing...

Tapping the artwork reveals or hides specific controls, and the ← button takes you back to your listings view. Most of the controls are pretty intuitive and need little elaboration, but for those of you who have just landed from Mars, here's a run-through of what's on offer.

> → **TIP** While browsing your music collection, any list of songs will include a Shuffle button at the top (⤨). Click to start a random selection of the current list.

• **Pause/Play a song** Tap ❚❚ and ▶ respectively, or press the mic button on the iPhone headset.

• **To skip** to the start of the current or next song, tap ◄◄ or ►►❙. In a podcast or audiobooks these buttons skip between chapters. You can also skip forward by quickly pressing the mic button on the iPhone headset twice.

• **To see the track list** of all the songs of the current album, tap the ▤ button. Tap the artwork preview, top-right, to get back to the Now Playing screen.

• **Scrubbing** To rewind or fast-forward within a song, slide the progress dot on the "scrubber" bar to the left or right. Alternatively, press and hold the ◄◄ and ►►| controls.

> ➔ **TIP** Slide your finger up and down the screen to adjust the rate of the scrubbing; this feature works for video playback too.

• **To adjust the volume** drag the lower onscreen slider to the left and right, or use the physical buttons on the side of the iPhone.

• **Stream music with AirPlay** Tap the ▭ icon to switch playback between your iPhone's speakers and any remote speakers connected to an Apple TV, Airport Express unit or other device that supports AirPlay. Find out more at apple.com/itunes/airplay.

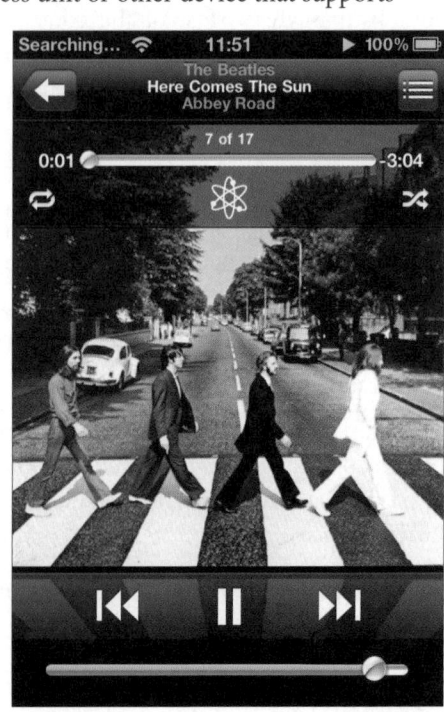

• **Shuffle** Tap the ⤬ icon to turn the shuffle selection feature on (blue) or off (white).

• **Repeat** The Music app offers two repeat modes, which are available via the ⟲ icon. Tap it once (it goes blue) to play the current selection of songs round and round forever. Tapping it again (⟲) repeats just the current track.

Cover Flow

A few years ago, Cover Flow was a standalone application, created by independent programmer Jonathan del Strother, that iTunes users could download as an alternative way to browse their music libraries. Apple liked it so much they bought the technology and incorporated it into the iTunes application, the iPhone, iPod Touch, and even their Mac operating system (where it can be used for file browsing).

Cover Flow is a slick-looking graphical interface for "flicking through" your music's album artwork, so it's perfectly suited to the iPhone, where you really can flick. It's like rooting around the record bins of your favourite music store.

To switch to the Cover Flow view on the iPhone, simply rotate the device through ninety degrees. The built-in accelerometer recognizes the shift of axis and displays your music library by its artwork. To view the track listing of an album, either tap the relevant image or hit **❸**. Then tap any of the songs in the list to set it playing.

The only disadvantage of Cover Flow mode is that you don't have access to Shuffle, Repeat or Ratings.

• **Genius** This feature generates a list of songs from your collection based upon accumulated iTunes Store information. In short, the Genius algorithms recognize that, for example, people who like The Beatles may well also like The Rolling Stones. Tap the ❋ icon to start – this will base the list on the currently playing song, or, if nothing is playing, prompt you to choose a song to define the list.

> **➜ TIP** You can have your iPhone play music, podcasts or videos for a certain amount of time and then switch off – like the sleep function of an alarm clock. Tap Clock and then Timer, and choose a number of minutes or hours. Then tap When Timer Ends, choose Sleep iPod, and hit Start.

Music settings on the iPhone

The iPhone offers various options for audio playback. You'll find these by clicking Settings on the Home Screen and then scrolling down to Music.

• **Shake to shuffle** With this feature turned on, shaking your iPhone while listening to the Music app will jump you to a random track in the current list (enabling shuffle if it wasn't already on). A source of much confusion, those who don't know of its existence frequently mistake this feature for a fault with their phone.

• **Sound Check** This feature enables the iPhone to play all tracks at a similar volume level so that none sound either too quiet or too loud. Because these automatic volume adjustments are pulled across from iTunes, the Sound Check feature also has to be enabled within iTunes on your computer. To do this, launch iTunes, open Preferences and under the Playback tab tick the Sound Check box.

• **EQ** Lets you assign an equalizer preset to suit your music and earphones. Note that you can also assign EQ settings to individual tracks in iTunes.

> **→ TIP** For more sophisticated equalizer settings try playing your iPhone music via the impressive EQu app, available in the App Store.

• **Volume Limiter** Lets you put a cap on the volume level of the iPhone's audio playback (including audio from videos), to remove the risk that you might damage your ears or indeed your earphones. Tap Volume Limit and drag the slider to the left or right to adjust the maximum volume level. If you're a parent, you might also want to tap Lock Volume Limit and assign a combination code to prevent your kids from upping the volume level without your permission.

• **Lyrics & Podcast Info** When enabled, the Music app displays lyric and episode information over the top of the Now Playing artwork screen. This will not work for all files as the data has to be present within the audio file. It generally, however, works for songs downloaded from the iTunes Store.

• **Group By Album Artist** A very useful setting for stopping complex "artist" listings set for specific tracks (say, where a song might "feature" someone else) fracturing the browsing experience of the "Artists" list in the Music app. Of course, for this to work, you need to make sure that you are making use of the "Album Artist" field in iTunes when prepping your library.

• **Create a playlist** Tap Playlists > Add Playlist… and follow the prompts, using the ⊕ icons to add individual songs. To edit playlists, select one to view its contents, tap Edit, and then use the ⊖ icons to remove items, and the draggable ≡ icons to change the order of the list. There are also buttons to Clear the playlist's contents, or Delete it completely.

> ➜ **TIP** Playlists you create on the iPhone will be moved across to your iTunes library next time you sync.

• **Star ratings** From the "Now Playing" screen, tap the ▤ icon and then slide your finger across the row of dots below the scrubber bar to add a rating for the currently playing selection.

• **Home button controls** Once music is playing, you can exit the Music app and continue to listen while using other apps. When the screen is locked, double-tapping the Home button reveals the Music play controls. When using other apps, double-tapping the Home button opens the Multitasking bar (see p.72) where Music controls can be found by swiping to the right.

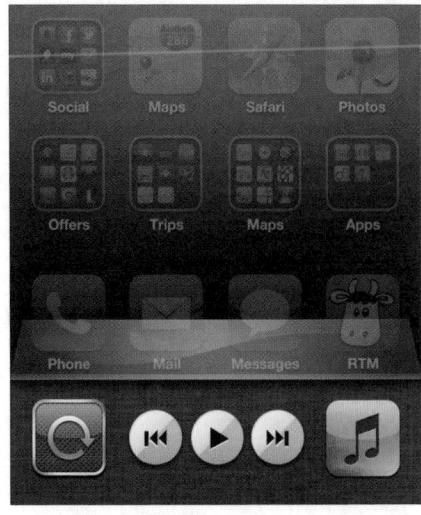

> ➜ **TIP** Sync the music on your iPhone from iTunes using a Smart Playlist of five-star ratings; you can then swiftly remove tracks you don't want by changing their rating on the iPhone.

Deleting music from an iPhone

You can't delete unwanted music directly from an iPhone. Instead, simply delete the track in iTunes and it will be deleted from your iPhone next time you connect and sync. Alternatively …

• If you want the music on iTunes but not on your iPhone, uncheck the little box next to the names of the offending tracks, and in the iPhone syncing options, choose "Only update checked songs".

• If you don't want to uncheck the songs, since this will also stop them playing in iTunes when in Shuffle mode, sync your iPhone with a specific playlist and remove the offending songs from that playlist.

• If you have Manual Music Management turned on (see p.45) simply connect your iPhone to iTunes and browse its contents via the iTunes sidebar (pictured here), deleting songs just as you would from a playlist by right-clicking and choosing the option from the dropdown menu.

> **➜ TIP** In the case of videos, you can delete them directly from the iPhone to free up space – simply swipe across and then tap Delete to confirm.

The spoken word

If the spoken word is your thing, then it is well worth checking out the Audiobooks and Podcasts sections of the iTunes Store (from either a computer or your phone) to see what's on offer.

Syncing with iTunes is carried out in pretty much the same way as it is for other types of content – you connect your iPhone to iTunes and then check the relevant boxes under the relevant tabs and hit Apply. Podcasts have their own tab, but audiobooks are hidden in the lower section of the Books tab in iTunes and can be easy to miss.

As for listening to your audiobooks and podcasts, the Music app is the place to go on the iPhone. Playback works the same as with music, although you have a couple of different options available to you from the Now Playing screen:

• **Tell a friend (podcasts only)** Tap the ✉ icon to send an email link to a friend so that they can find the podcast you are listening to in the iTunes Store.

• **Backtrack 30 seconds** Tapping the "30" icon, will rewind the audiobook or podcast by thirty seconds.

• **Playback speed** To adjust the playback speed of an audiobook or audio podcast, tap the "1x" icon to the right of the scrubber bar (to choose either "2x" or "½x").

Remote controls

The iPhone is great as a remote control for any number of audio – and video – setups. Here are a few apps from the App Store that will help you get the job done:

Apple Remote

Free to download from the App Store, Apple's Remote app can be used to control iTunes on your computer via Wi-Fi. Coupled with an Apple Airport Express unit, this can be a great way to stream music using AirPlay from a computer in the bedroom, say, to your hi-fi system in the living room. It also works with an Apple TV unit connected to your television, allowing you to move around the onscreen menu system by swiping and tapping on the iPad's screen. For the full story on the Remote app, visit apple.com/itunes/remote, and for more on the Apple TV, point your browser at apple.com/appletv.

Rowmote Pro

Though it'll cost you a few dollars, this super-charged remote control app is well worth having a play with, as it gives you control of all sorts of applications on your Mac computer, not just iTunes.

VLC Remote

If you use the excellent VLC player on your computer for watching movies, this remote control app is well worth a couple of dollars.

→ **TIP** There are also loads of third-party remotes in the App Store for controlling Spotify playback.

Apps: Music

For fans, for musicians

For many people, owning an iPhone means the end of their relationship with their trusty iPod. As a multifunctional device, the iPhone is capable of so much more than its MP3-playing predecessor, so it is worth exploring some of the other audio-related tools available to you. In this chapter we pick out a few of the best.

Apps for listening to music

Aside from the iPhone's built-in Music app, there are plenty of other apps that'll let you listen to music… assuming you can connect to the Internet.

Spotify

With this app and a Premium Spotify account, you can stream unlimited

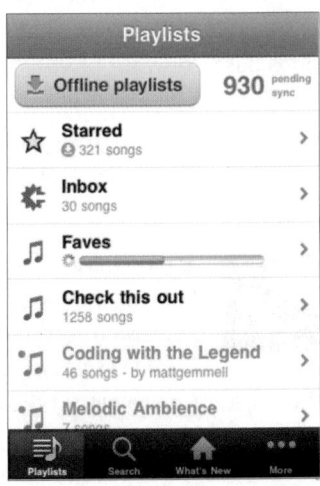

music to your iPhone from the Internet each month (and also play it offline) for around the price of a CD album. Though not yet available in some countries, it's pretty compelling for those who can get it.

AccuRadio

There are plenty of Internet radio players to be found in the App Store; this one is free, has an easy-to-use interface, and lets you create favourites lists.

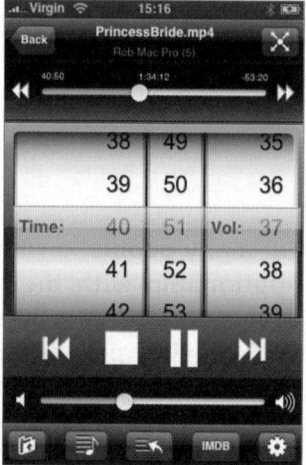

TuneIn Radio

Another great radio tuner app. Though not free, it's still a bargain, given that it offers more than 40,000 stations.

Pandora

Only available in the US, this well-established Internet radio service has the added twist of personal recommendations based on your taste.

> ➡ **TIP** To instantly identify songs using the iPhone's microphone, try the SoundHound or Shazam apps.

VinylLove Pocket

For the nostalgic among you, this virtual turntable lets you listen to the music on your iPhone with a little added crackle. You also get to move the turntable's arm to choose the track, or place within the track, that you want to listen to.

Apps for making music

The Music category within the App Store is awash with virtual instruments, sequencers, drum pads and other noisy creations that, in the hands of most people, will make you want to stuff cotton wool into your ears. With a bit of perseverance and a sprinkling of talent, however, there are some that can be made to sound more than just a novelty. Here's a few worth looking at:

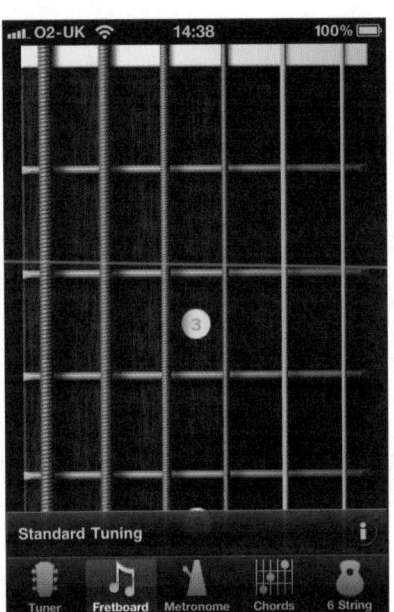

GuitarToolKit

When the App Store was first launched the world was wowed by virtual guitar apps. A few years on and there are hundreds available. This is one of the better ones, complete with a tuner and metronome to help you play one of those old-skool wooden versions.

Looptastic Producer

An amazing dance-music creation tool, with loads of built-in loops and the ability to add your own. There's also a bunch of effects and time-stretch tools.

NLog MIDI Synth

Arguably the best Korg-styled tone generator for the iPhone, with a delightful analogue feel to the interface.

Nota

A fantastic music theory app, featuring a piano chord and scale browser, a landscape keyboard for practicing, and a quiz for testing your progress.

➜ **TIP** Remember that the headphone jack can be used as a line-out to play your creations via a PA, amplifier or hi-fi system instead of the iPhone's built-in speaker.

Songsterr Plus

More guitar and drum kit tabs laid out in one place than you could shake a stick at.

Songify

This one has been a real online sensation. You basically sing a few lines into your phone and the app instantaneously transforms it into a tuneful, groovy pop hit.

FourTrack

This is an incredible multitracking mini recording studio. It can be a bit fiddly to use, but is also very useful for quickly getting song ideas down.

➜ **TIP** The iPhone also comes with a built-in audio recorder called Voice Memos, which automatically transfers recordings you make back to iTunes when you sync.

iStylophone

We can't be sure if Rolf Harris has an iPhone, but if he does, this app almost certainly gets a regular airing.

19

iPhone video

Movies, TV shows, YouTube

There's nothing particularly revolutionary about a phone or portable media player being able to handle video files. What's different about the iPhone is the size and clarity of its screen, the user-friendly controls and the decent battery life. A new iPhone will play video for around six or seven hours before running out of juice – more than enough for a couple of average feature films.

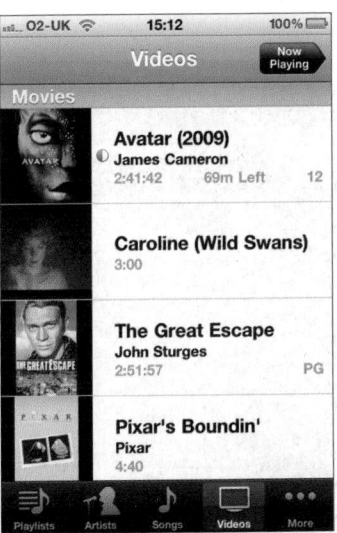

Playing videos

To watch videos on the iPhone, head to the Videos app and start browsing your lists. You can also drag to the top of the list to reveal a search field. Depending on the content you have either downloaded from the iTunes Store or synced across from iTunes, you'll find the list divided up into categories: Movies, TV Shows, Podcasts and Music Videos. A blue dot next to a TV Show in the Videos list on the iPhone means that the episode has not yet been viewed. Tap an item to start playback. Once the video is playing, tapping the screen reveals play, volume and scrubbing controls, just like for audio. Additionally, you can:

• **Toggle views** Tap the ▪ and ▪ icons to toggle between theatrical widescreen and full-screen (cropped) views

> → **TIP** You can also toggle between the two view modes by simply double-tapping the screen.

• **Display subtitles** Where the file you are watching supports them, tap the "speech bubble" icon to the left of the play controls to access audio and subtitles options for the playing movie.

• **Display chapters** Where the file you are watching supports them, tap the ▤ icon to the right of the play controls to display the chapters of the movie you are watching.

To delete a movie from the iPhone (which can be really handy if you want to make space for more content when out and about), simply swipe across its entry and then tap Delete to confirm.

Video out

The iPhone can output both NTSC and PAL TV signals so that you can connect your iPhone to a TV or projector. This can be done using one of the following, which all attach to the iPhone via the Dock connector:

• **Apple iPhone Dock Connector to VGA Adapter** This is great for connecting to most standard projectors.

• **Apple iPhone Dock Connector to HDMI Adapter** This is useful for connecting to most modern TV and projector setups.

• **Apple Component AV Cable** This gives you five RCA-style plugs: three for the video and two for the audio.

• **Apple Composite AV Cable** This gives you three RCA-style plugs: one for the video and two for the audio.

These accessories need to be purchased separately from Apple. Many third-party manufacturers make cheaper versions, which should work with the iPhone, assuming they have the appropriate Dock connector, though it's always worth checking with the manufacturer before you buy, as some may only work with older iPods or iPhones.

AirPlay and screen mirroring

AirPlay is an Apple tech-
nology that allows you to
stream content from an
iPhone to an Apple TV,
connected via an HDMI
cable to either a TV set
or projector. To get this
working while using the
Videos app (or any other
AirPlay enabled app),
tap the 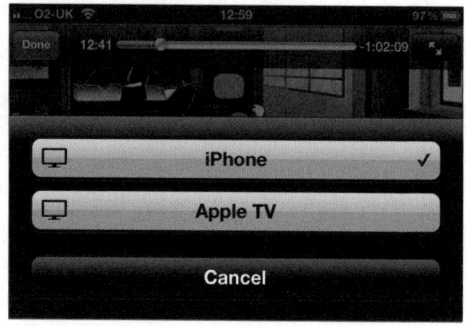 icon to switch
playback between your iPhone's screen and an Apple TV on the same
Wi-Fi network.

If you have an iPhone 4S, it is possible to "mirror" whatever you
happen to have on your iPhone's screen at any one time to an Apple TV
– even if the application in question is not specifically AirPlay-enabled.
This is great for everything from playing iPhone games on a bigger
screen to quickly zapping a webpage to the TV so that the whole family
can see it. To start mirroring, double-tap the Home button to reveal the
Multitasking tray and then swipe to the right so that you can see the
player controls. Next tap the  icon, choose the Apple TV you want
to stream to and then slide the Mirroring switch to the On position.

To find out more about Apple TV, visit apple.com/appletv.

YouTube

The iPhone comes with a dedicated YouTube app, which can be used to access all the content from YouTube (assuming you are connected to the Internet). Once a clip is playing, tap to see the onscreen playback controls. They work in exactly the same way as the video controls in the Videos app (see p.184), though here you can additionally:

• **Create Favorites** Tap the ☐ icon to add a video clip to your Favorites list.

• **Share clips** Tap the ☒ icon to send a link by email.

You can also sign in using a YouTube account login (Google account logins also work) allowing you to access your uploads, subscriptions, favourites, and also:

• **Leave feedback** Tap the blue ◉ icon next to a clip and scroll down to find the option to Comment or Rate (out of five), or Flag (if you find the clip offensive).

• **Create playlists** Tap More > Playlists to edit, create and delete YouTube playlists.

> ➔ **TIP** If you use YouTube a lot, you might want to edit which icons appear on the category strip at the bottom of the screen. To do this tap More > Edit and then drag and drop the icons you want into position.

20
Apps: Video

Streaming TV and other video players

Sure, you can purchase pretty much any TV season you fancy from the iTunes Store, but there are also many streaming TV services that work on the iPhone (even though it lacks Flash). There are also a number of excellent video player apps for the iPhone that give you access to a lot more video formats than Apple's Videos app allows straight out of the box.

Streaming live TV

The main limitations to watching either live or catch-up TV on the iPhone are regional: many services that are currently accessible in Europe aren't in the US, and vice versa. You are sure to find some that work, but don't be too surprised if a few listed below dish you up a whole lot of nothing.

Where a web address appears after the name in the lists below, the recommendation is an optimized website, or "webapp", that can be viewed in Safari and added to your iPhone's Home screen as a Webclip by tapping the 🖻 icon and choosing Add to Home Screen.

iPlayer (bbc.co.uk/mobile/iplayer)

Until the BBC get around to releasing a dedicated iPhone app for viewing their UK catch-up service, you'll have to make do with this Safari-optimized website.

TVCatchup (tvcatchup.com)

This website is the only thing you need to stream live UK TV to the iPhone for free through Safari. It isn't great over a cellular data connection, but works very well over Wi-Fi.

Recording from TV

If you want to create iPhone-friendly videos by recording from television, your best bet is to use a TV receiver for your computer. Some high-end PCs have these built in, but if yours doesn't you should be able to pick one up relatively inexpensively, and attach it to your PC or Mac via USB.

The obvious choice for Mac users is Elgato's superb EyeTV range of portable TV receivers, some of which are as small as a box of matches. You can either connect one to a proper TV aerial or, in areas of strong signal, just attach the tiny aerial that comes with the device.

With an EyeTV, it's easy to record TV shows and then export them directly into an iPhone-friendly format. They also produce a very good app that allows you to stream your recordings straight to your iPhone, check the TV schedules and even set recordings going on your EyeTV box at home when out and about. For more info, see:

Elgato elgato.com

Hauppauge, Freecom and various other manufacturers produce similar products for PC owners. Browse Amazon or another major technology retailer to see what's on offer. If you can't find one that offers iPhone/iPod export functions, record the shows in any format of your choice and use QuickTime Pro to re-save them for the iPhone.

UK TV & Radio

This dedicated app does a similar job to the previous two webapps. It offers a familiar app experience that the browser-based services don't, but you will pay for the privilege.

TVUplayer

This app pushes out more than 900 live TV channels from around the world.

Boxee

Boxee are one of the big names in web-based streaming TV services and set-top box systems. At the time of writing, the Boxee app is iPad-only, but it is hoped that an iPhone version is in the pipeline. For their latest news, visit boxee.tv.

Other video players

The biggest limitation of the iPhone when it comes to video playback is the limited number of file formats that are supported by the Videos app. Thankfully, there are several apps available in the store that offer an alternative:

GoodPlayer

Great all-round video player for the iPhone that can handle pretty much any file format that you throw at it (Xvid, Divx, MKV, MP4, etc).

ProPlayer

Another good option that can handle all manner of files. Additionally, ProPlayer has loads of additional features and tools for organizing and playing your files. There are also some nice Multi-Touch gestures that make playback very intuitive.

Reading

iBooks

eBooks on the iPhone

Despite its small screen, the iPhone makes a surprisingly good eBook reader – especially since the launch of Apple's iBook app and iBookstore, which makes thousands of books available at the touch of a button.

iBooks is not one of the default apps on the iPhone. It is a free app, but you will have to visit the App Store to download it. You might even be prompted to download it the first time you turn on your phone.

Unfortunately, not every country has an iBookstore, so if you can't find the iBooks app in your local App Store, it basically means that you are going to have to find another way of getting books onto your iPhone. (Skip forward to the next chapter to find the answer.)

Using the iBooks app

Once launched, iBooks displays either a pretty bookshelf (your Library) or the iBookstore (which looks very similar to the iTunes Store). To toggle between the two, tap the button in the top-right corner of the app. When viewing your bookshelf you can switch between views of Books, PDFs, or new categories of your own making by tapping the Collections button. Though you can't purchase PDFs from the iBookstore, they can be synced across from iTunes or moved to the app from an email attachment.

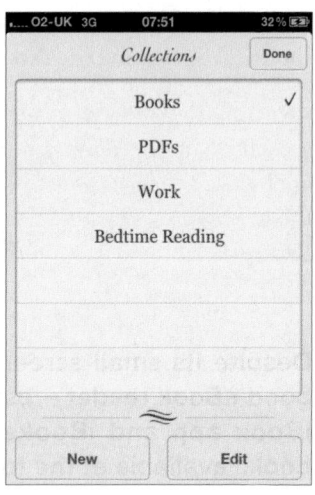

> ➔ **TIP** You can toggle between the "bookshelf" and "list" views of your library using the two buttons on the top-right.

To change the position of a book on your shelf, tap and hold until it appears to lift, and then drag. To either delete books or move them to a different Collection, tap Edit, then tap the item or items in question and then either Move or Delete.

Using the iBookstore

There really aren't any surprises here. You browse books and purchase them in just the same way that you do music, movies and TV shows in the iTunes Store. You can even use the same login details. Start exploring and you'll quickly get the hang of it.

• **Sample chapters** In most cases you can tap Get Sample to download a free excerpt to whet your appetite. You can download the full text at any time while reading (assuming you are connected to the Internet) by tapping the Buy button at the top of any page.

• **Redownloading purchases** Tap Purchases in the iBookstore to see a list of all the books you have previously downloaded. If there are items in the list that are no longer on your iPhone, and assuming the books are still available in the store, you should see options to Redownload them again for free.

Reading with iBooks

To open a book (as opposed to a PDF), simply tap the Books tab within your iBooks Library and then choose the title you want to read. To turn a page, either drag the bottom corner, or tap to the left or right of the text near the edge of the iPhone's screen.

A single tap anywhere on the body of the page will reveal or hide further options at the top and bottom, including brightness, a bookmark,

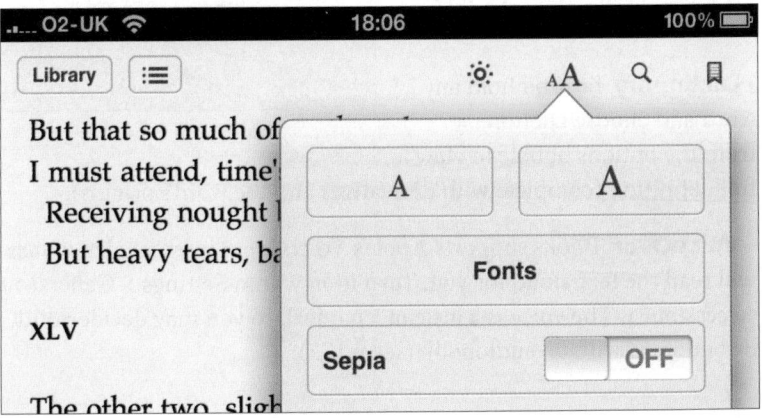

text options and text search. This tap also reveals a slider at the bottom of the page to help you quickly jump to another part of the book.

> **➔ TIP** Look within Settings > iBooks to determine whether tapping on the left margin takes you to the next page or the previous page. Also note the options for text justification and hyphenation.

• **Contents page** Tap the ▤ button to view the Contents page of the title you are reading; once there, tap Resume to return to the point where you were reading.

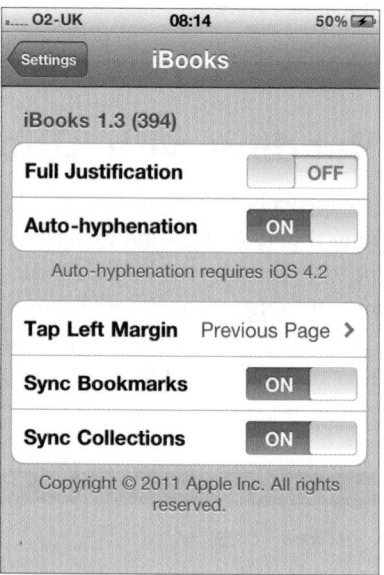

• **Search** Tap and hold any word and choose Search from the options bubble; this displays a tappable list of other places where the word occurs and links to search Google and Wikipedia (both of which take you out of iBooks and into Safari).

• **Dictionary** Tap and hold any word and choose Dictionary from the options bubble to view a full definition (complete with derivatives and the word's origins).

• **VoiceOver** iBooks supports Apple's VoiceOver screen reader, which will read the text aloud for you. Turn it on within Settings > General > Accessibility. The voice is a little mechanical, so you may decide you'd be better off with an audiobook (see p.177).

Adding Notes, Highlights and Bookmarks

Tap and hold the text to which you want to add a note or highlight and then drag the blue anchor points to adjust the size of the selection. When you are ready, tap either Highlight or Note (and in the latter case, start typing). Tapping margin notes and highlighted text again reveals options to change colours and delete. To add a bookmark to a page (and you can have as many as you want), tap the  icon next to the search icon at the top.

To view a list of all the highlights, notes and bookmarks in a given book, tap the ☰ button at the top, and then tap Bookmarks. From here you can also delete specific items from the list by swiping and then tapping Delete.

> **→ TIP** iBooks remembers where you are when you exit a book and will take you to the same point next time you open the title.

Reading PDFs

When viewing a PDF sent as an attachment in an email (see p.121) or online within Safari (see p.218), you should see an "Open in iBooks" button which will add the PDF to your iBooks library. Once this is done, the file will become available in the PDF section of your bookshelf – even when you are no longer connected to the Internet.

The reader controls for PDFs are very similar to those for books, but without the specific text options. If you don't get on with iBooks as a PDF reader, try one of the alternative apps mentioned in the next chapter.

Deleting books and PDFs

To delete books from your iPhone's Library, tap Edit, tap to select the titles you want to ditch, and then tap Delete to confirm. Books purchased from the iBookstore will still be available to sync back onto the iPhone later for free via the Purchased tab in the store, should you wish to do so.

Syncing with iTunes

You can build up a sizeable collection of ePub files on your iPhone without having to worry about storage space. Text-only ePub books, even long ones, will only be a couple of megabytes at most (that's about the size of an average digital photo file); ePub books and PDF documents with lots of images, on the other hand, can be quite chunky.

> → **TIP** To keep an eye on how much space books are taking up on your iPhone, connect to iTunes and look at the multicoloured status strip at the bottom of the Summary panel.

To start syncing your books and PDF documents with iTunes, connect your iPhone, select the Books tab, and then check the appropriate sync options. Note that you can choose to sync just books, just PDFs or both. Once this is set up, new iBookstore downloads and stored PDFs will be synced back to iTunes every time you connect.

Even though you can't actually read your synced iBookstore purchases within iTunes on your computer, it's still a worthwhile exercise as a means of creating a backup of the titles you have downloaded to the iPhone. PDF files, on the other hand, can be dragged out of iTunes on your computer to be read onscreen.

Also, on your iPhone, head to Settings > iBooks to set options for whether you want any Bookmarks or organized Collections to be synced back into iTunes along with the books themselves.

ePub books from other stores

iBooks can also display books in the ePub format from sources other than the iBookstore (assuming they don't have any special DRM copy protection built in). First, you need to get them into iTunes on your computer. Highlight the Books listing in the iTunes sidebar and then drag and drop the files into the main iTunes window, where they will appear alongside your iBookstore purchases. You can then connect your iPhone and sync them across in the normal way.

This won't work for Sony Reader Store titles (which employ a form of DRM), but there are plenty of other ePub sources that you can turn to, including:

Google Books books.google.com
ePubBooks epubbooks.com

Whether you are an author yourself, or a student, there are loads of benefits to being able to create your own ePub files. To give it a go, try:

Storyist storyist.com
eCub juliansmart.com/ecub

22

Apps: Other ways to read

Readers, comics and kids' books

Using iBooks is not the only way to read eBooks on the iPhone. And when it comes to specialist graphic titles, such as comics and manga, you are far better off turning to the App Store than the iBookstore to get your fix.

eBook reading apps

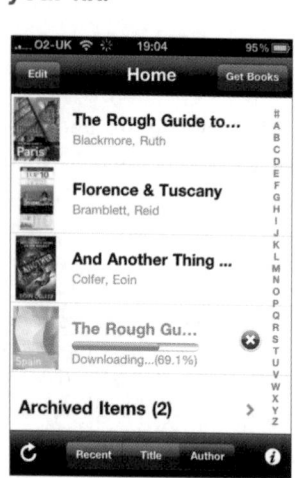

Kindle

For reading Amazon eBooks this a great little app and has a similar set of features to iBooks. You can't actually browse the Amazon store from within the app; for that you need to head to the website and then sync your purchases over the airwaves.

Stanza

This reader, and its associated store, offer thousands of titles and you can also add your own files using the Stanza Desktop application.

Google Books

Syncs via the cloud with your Google Books account and has a nice reader interface which, like Stanza, includes a white-on-black night-reading mode. As with the Kindle app, you have to shop for titles outside of the app: ebooks.google.com.

Kobo

A nice-looking app that gives you access to thousands of titles from the Kobo store. The reading experience is excellent, with font and bookmarking options.

Kids' reading apps

There is a glut of kids' content to be found in the App Store Books category, but do read the reviews and be selective, as the majority of what's there is shockingly poor. Here are a few that are worth a go:

Ladybird Classic Me Books

This a beautiful reader app that gives you access to the iconic Ladybird Me Books series. Many feature audio of the text read by famous names, and you can even record your own sound effects and narrative as you read.

The Cat in the Hat

Ingeniously adapted for the screen, you can read it like a book, listen to the narrative, or play the whole thing through like a movie.

DK My First Words

A great introduction to iPhone reading for the very young. Loads of letter-based games and a beautifully rendered paint-and-reveal feature.

Comics and graphic novels

Marvel

Access to an essential store of classic superhero comics. Download and read either page by page or frame by frame.

iVerse Comics+

A really easy-to-use store and comic reader. There are both paid-for and free comics to be found here.

iMangaX

More than a thousand manga. All the content is stored online, so an Internet connection is essential.

IDW Comics

The complete IDW collection available in one app. Like the other stores, you buy individual comics and then organize your own library.

➜ **TIP** As well as all the "store" and "library" apps, there are also thousands of comics to be found in the iTunes App Store published as standalone apps.

23

Apps: News & RSS

Making the most of news on the iPhone

The iPhone is without question a device more useful for consuming than creating. As such, it's ideally suited to delivering your daily fix of current affairs, or any other kind of news for that matter, wherever you want it, and in a perfectly digestible format. So, whether you read in bed, on the train, on the couch or at the breakfast table, the iPhone is a handy replacement for a newspaper, magazine or periodical.

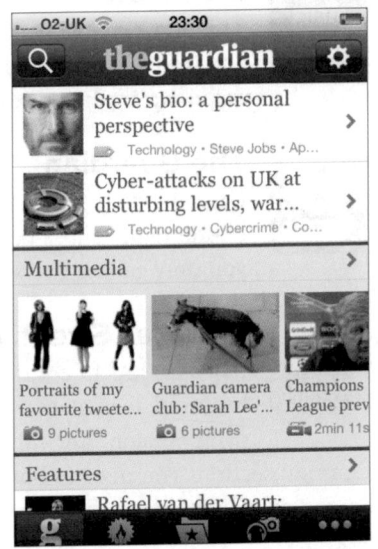

News apps

Many of the major news and magazine publishers are battling to define their place on the iPhone, and within the new digital marketplace as a whole. Some media companies have opted to make their apps available for free, but with advertising support; some are trying to get users to pay a subscription; others charge a one-time fee for users to download their app.

The real question here is not whether you want to pay for your news (there are still a thousand places online where the latest stories can be harvested for free), but whether you are prepared to pay for a particular stance, attitude or editorial voice – which is what we do in the real world when we choose to hand over cash for a print newspaper.

All the major news voices are represented on the iPhone, and there is even a News category in the App Store to help find what you are looking for. In terms of useability, here are a couple of recommendations:

BBC News

This app has a really nice split-screen view in both landscape and portrait modes, complete with a handy news-ticker that dishes up the latest headlines.

The Guardian

Great-looking subscription UK news app with loads of photography and a customizable download feature, so that you only get the news categories you want.

The Wall Street Journal

Register for free to see the latest news presented in a beautiful format that successfully mimics the look of the print version. For full access to the Business, Markets and Opinion sections you need to buy a subscription.

USA Today

A really nice interface with an easy-to-navigate layout. Scrolling from page to page in long articles is particularly well handled.

Pulse

As one of the more beautiful RSS aggregators (see p.206), Pulse pulls together newsfeeds from multiple sites to create a mosaic of content that is easy and fun to digest.

Zinio

This app is a one-stop shop for magazines. There are hundreds on offer and plenty of free samples, and you can choose to purchase either single issues or annual subscriptions. There's a handy calendar view that shows you which issues you'll be getting to read each month.

> **→ TIP** The Financial Times no longer has a dedicated app in the App Store; instead you need to visit their excellent webapp (app. ft.com). It is free to browse a small amount of teaser content, but you have to sign up for a subscription to get the full package.

Newsstand

Tap the special Newsstand icon on your Home Screen to get quick access to subscription magazine and news services that publish into the Newsstand category of the App Store. To furnish your iPhone's Newsstand with publications, either browse via the App Store or tap the shortcut button at the top of the expanded Newsstand panel. In most cases, you can then subscribe or purchase new issues from within each publication.

RSS feeds

RSS – Really Simple Syndication – allows you to view "feeds" or "newsfeeds" from blogs, news services and other websites. Each feed consists of headlines and summaries of new or updated articles. If you see something you think you'd like to read, click on the headline to view the full story. You can use a tool called an aggregator or feed reader to combine the feeds from all your favourite sites. It's almost like having your own personalized magazine or newspaper.

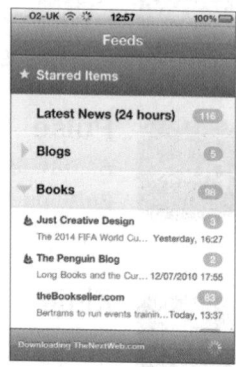

At the time of writing, arguably the best web-based aggregator is Google Reader (reader. google.com), which can be accessed via Safari. To get started, simply sign in using the same credentials you use for other Google services and start adding feeds.

The alternative on the iPhone is to download a dedicated RSS feed app, such as Feeddler, Early Edition, NetNewsWire (pictured) or NewsRack. Search for them in the iTunes App Store.

Reading List

Also worth investigating is the Reading List feature built into Safari on the iPhone. When you find a webpage you'd like to read later, tap the 🖅 button and choose "Add to Reading List". You'll find the Reading List at the top of your Bookmarks menu, accessed via the ꁈ button in Safari.

If you don't get on with Apple's Reading List, sign up for the excellent Instapaper service (instapaper.com) and app.

> **→ TIP** If enabled within Settings > iCloud, your Reading List will sync with your other Apple devices along with Bookmarks.

The
Internet

24

The web & Safari

Browsing on the iPhone

The iPhone certainly isn't the first mobile device to offer web browsing, but, arguably, it's the first one to provide tools that make it a pleasure as opposed to a headache. As discussed earlier, the iPhone comes with a nearly full-fledged version of the Safari web browser.

The basics

Make sure you have a carrier data signal (or, even better, a Wi-Fi signal), and tap Safari on the Home Screen. Then…

• **Enter an address** Click into the address field at the top of the screen, tap ⊗ to clear the current address and start typing. Note the ".com" key (tap and hold it to reveal alternatives).

• **Search the web** Tap into the search field at the top of the screen and start typing. When you're done, hit Search. You can select any of the suggested search items that appear while you're typing.

→ **TIP** If you want to switch from Google (the default) to Yahoo! or Bing searching, look within Settings > Safari > Search Engine. Of course, you can also visit any search engine manually and use it in the normal way. For more search tips, see p.216.

• **Search the current page** As you type into the search field, also note that below the list of suggested web search results there is a link for viewing search results found on the currently loaded page. Tap this and then use the ◀ and ▶ buttons to skip between the various instances of the phrase or word.

• **To follow a link** Tap once. If you did it by accident, press ◀.

→ **TIP** You can see the full URL of any link by tapping and holding the relevant text or image. (This is equivalent to hovering over a link with a mouse and looking at the status bar in a normal web browser.)

History & cache

Like most browsers, Safari on the iPhone stores a list of each website you visit. These allow the iPhone to offer suggestions when you're typing an address but can also be browsed – useful if you need to find a site for the second time but can't remember its address. To browse your history, look at the top of your Bookmarks list, accessible at any time via the ⌘ icon. To clear your history, either tap Clear at the bottom of the screen or look for the option in Settings > Safari.

Unfortunately, despite storing your history, Safari doesn't "cache" (temporarily save) each page you visit in any useful way. This is a shame, as it means you can't quickly visit a bunch of pages for browsing when you've got no mobile or Wi-Fi reception. It also explains why using the Back button is slower on the iPhone than on a computer – when you click ◀, you download the page in question afresh rather than returning to a cached version. If you need to download pages to read later, offline, try an app such as Instapaper or the appropriately titled Read It Later.

- **Reload/refresh** If a page hasn't loaded properly, or you want to make sure you're viewing the latest version of the page, click ↻.

- **To share a page's address** When viewing a page, tap the ➦ icon and then tap Mail Link to this Page. A new email will appear with the link in the body and the webpage's title in the subject line. You can also tap Tweet to publish the address to Twitter. For this to work you must first enter your Twitter account credentials within Settings > Twitter.

- **Zoom** Double-tap on any part of a page – a column, headline or picture, say – to zoom in on it or to zoom back out. Alternatively, "pinch out" with your finger and thumb (or any two digits of your choice). Once zoomed, you can drag the page around with one finger.

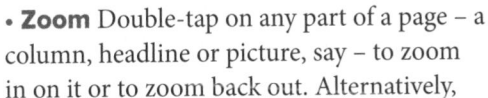

➡ **TIP** If you struggle with the onscreen keyboard, try rotating the phone when typing – landscape mode offers bigger keys.

- **To open a page in Reader view** Reader view gives you a clean, ad-free text-and-pictures version of the current page, allowing you to read articles and posts in a far more digestible iPhone-friendly format than most websites offer. There is even a link at the top of the page to change the text size (similar to the feature found in iBooks). If Reader view is available for the page you are looking at, a Reader button will appear in the address field at the top of the screen.

Multiple pages

Just like a browser on a Mac or PC, Safari on the iPhone can handle multiple pages at once. These are especially useful when you're struggling with a slow connection, and you don't want to close a page that you may want to come back to later.

• **To open a link in a new page** Tap and hold any link (either text or an image) and choose the option from the panel that pops up. You can also determine within Safari > Settings whether you want to view new pages straight away (Open Links > In New Page) or have them opened out of sight (Open Links > In The Background).

• **Open a blank new page** Tap ⧉, then New Page.

• **Switch between pages** Tap ⧉ and flick left or right. To close a page, tap ⊗.

Bookmarks

Bookmarks, like Home Screen webclips (see p.73) are always handy, but when using a device without a mouse and keyboard, they're even handier than usual. To bookmark a page to return to later on the iPhone, tap ⎙ and then Add Bookmark. To retrieve a bookmark, tap ⬚, browse and then click the relevant entry. To edit your bookmarks…

• **To delete a bookmark or folder** Tap Edit, followed by the relevant ⊖ icon. Hit Delete to confirm.

• **To edit a bookmark or folder** Tap Edit, hit the relevant entry, and then type into the name and URL fields.

• **To create a new folder** Tap Edit, then New Folder.

• **To move a bookmark or folder** Tap Edit and slide it up or down using the ☰ icon. Alternatively, tap Edit, then hit the relevant entry and use the lower field to pick the folder you'd like to move the bookmark or folder into.

Syncing bookmarks from your Mac or PC

Using iCloud

The easiest way to sync your bookmarks with Safari on either a Mac or PC, or with Internet Explorer on a PC, is via iCloud. To check that bookmarks are set to sync, look within Settings > iCloud on your phone. Then, on a Mac go to System Preference > iCloud, and on a PC launch the iCloud Control Panel (icloud.com/icloudcontrolpanel) to check that the sync is set up from that end.

Using iTunes

You can also sync bookmarks via iTunes. Just connect your iPhone (either wirelessly or via a cable) and click its icon in iTunes. Click the Info tab and check the relevant box under Other. The bookmarks will move across to the iPhone once you have hit Apply and then Sync.

➔ **TIP** Use the free Xmarks service (xmarks.com) to sync other browsers' bookmarks to Safari, and in turn the iPhone.

Likewise, bookmarks from the iPhone will appear back in your browser (almost instantaneously if you are using iCloud).

➔ **TIP** The Bookmarks screen also gives you access to your Safari Reading List, where you can store links to articles or webpages you might want to come back to and read later (but not keep for posterity as Bookmarks).

Forms & AutoFill

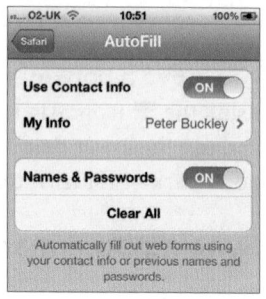

You can set Safari on the iPhone to remember the names and passwords that you use frequently on websites, though there is a risk that someone else could use those credentials if they get hold of your phone. You can turn the feature on or off within Settings > Safari > AutoFill – look for the option to turn on Names & Passwords. If you do enable it, also consider setting a four-digit screen unlock code as an extra level of security. This can be found within Settings > General > Passcode Lock.

AutoFill can also help you when filling out address fields on webpages, though you will need to tell the iPhone which contact details to use for this. First enable AutoFill, as already described, and then turn on Use Contact Info, choose My Info and select your contact entry from your iPhone's Contacts list.

→ TIP When typing into form fields on a webpage, use the Previous, Next and AutoFill buttons just above the keyboard to make best use of the AutoFill feature.

If at any time you wish to remove all the saved passwords and usernames on your iPhone, tap Clear All.

Safari settings

You'll find various browsing preferences under Settings > Safari. Here you can clear your history (useful if Safari keeps crashing, or you want to hide your tracks), play with security settings (make sure that the Fraud Warning option is switched on) and choose a search engine. You can also access the following:

• **Private Browsing** With this turned on, your Safari activity will not generate any stored browsing history or add any new entries to your AutoFill data.

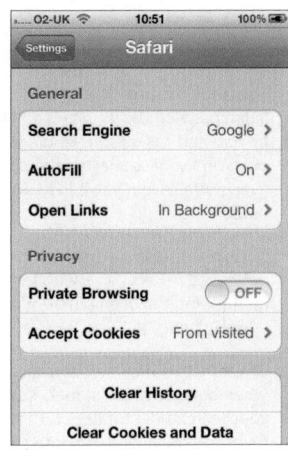

• **JavaScript** This is a ubiquitous way to add extra functions to websites and is best left on.

• **Pop-up blocker** Stops pop-up pages (mainly ads) from opening.

• **Cookies and data** Cookies are files that websites save on your iPhone to enable content and preferences tailored for you, for example, specific recommendations on a shopping site. To delete all this data, tap the Clear Cookies and Data button. Alternatively, tap through to the Advanced > Website Data screen to see exactly how much data specific sites are saving to your phone. From here you can tap Edit and then delete the cached data for specific sites if you so wish.

Webpage display problems

If a webpage looks weird on screen – bad spacing, images overlapping, etc – there are two likely causes. First, it could be that the page isn't properly "Web compliant". That is, it looked OK on the browser the designer tested it on (Internet Explorer, for example), but not on other browsers (such as Safari on the iPhone). The solution is to try viewing the page through another browser (see p.219, or use the Safari Reader feature, see p.211).

Second, it could be that the page includes elements based on technologies that the iPhone doesn't currently handle, such as Flash, Real and Windows Media. This is especially likely to be the problem if there's a gap in an otherwise normal page.

If a webpage looks OK, but different from the version you're used to seeing on your Mac or PC, it could be that the website in question has been set up to detect your browser and automatically offer you an optimized mobile version (see p.74).

iPhone Googling Tips

The Google search field at the top of each Safari page on the iPhone is an incredibly useful tool and as you start to type, its dynamic popover list of suggested searches makes it even more so. But you can make it work even harder.

Google also offer a very easy to use iPhone-optimized results page, with tabs at the top for jumping between different types of search – "web", "images", "local", etc.

Given that your connection speed may be low when out and about, it makes sense to hone your search skills for use on the iPhone. All the following tricks work on both a PC and Mac. Typing the text as shown here in bold will yield the following search results:

Basic searches

william lawes > the terms "william" and "lawes"

"william lawes" > the phrase "william lawes"

william OR lawes > either "william" or "lawes" or both

william -lawes > "william" but not "lawes".

All these commands can be mixed and doubled up. Hence:

"william lawes" OR "will lawes" -composer > either version of the name but not the word "composer".

Synonyms

~mac > "mac" and related words, such as "Apple" and "Macintosh".

Definitions

define:calabash > definitions from various sources for the word "calabash". You can also get definitions of a search term by clicking the definitions link at the right-hand end of the top blue strip on the results page.

Flexible phrases

"william * lawes" > "william john lawes", etc, as well as just "william lawes".

Search within a specific site

site:bbc.co.uk "jimmy white" > pages containing Jimmy White's name within the BBC website. This is often far more effective than using a site's internal search.

Search web addresses

"arms exports" inurl:gov > the phrase "arms exports" in the webpages with the term gov in the address (i.e. government websites).

Search page titles

train bristol in title:timetable > pages with "timetable" in their titles, and "train" and "bristol" anywhere in the page.

Number and price ranges

1972..1975 "snooker champions" > the term "snooker champions" and any number (or date) in the range 1972–1975.

$15..$30 "snooker cue" > the term "snooker cue" and any price that falls in the range $15–30.

Search specific file types

filetype:pdf climate change statistics > would find PDF documents (likely to be more "serious" reports than webpages) containing the terms "climate", "change" and "statistics".

Searching without your keyboard

To search the web without even typing, you can use Siri (see p.53) and start a search simply by saying "Search the web…" Alternatively, head to the App Store and download the Google Search app, which allows you to not only search by talking (Voice Search), but also by pointing your iPhone's camera at something you want to know about (Google Goggles) or based on where you are (My Location).

Viewing PDFs and Word documents

The iPhone can view Word, Excel and PDF documents on the web, creating an iPhone-friendly preview version in a popover window. Once opened, scroll down to read subsequent pages, and double-tap to zoom in just as you would with a regular webpage.

In the case of PDFs, assuming you have the free Apple iBooks app installed (see p.193), look out for the link to "Open in iBooks" (tap the screen once if you don't see it). This link will save a copy of the PDF to your iBooks bookshelf. The file is then synced back to iTunes next time you sync either wirelessly or via a cable.

If you use a different PDF reading app, such as GoodReader, or want to move the PDF to a file storage app (such as Dropbox), tap the "Open in…" link and choose your app from the list.

→ **TIP** If you're following a link from Google to a PDF, Word or Excel doc, and it is taking an age to download, click the "View As HTML" link instead of the main link to the document. This way you'll get a faster-loading text-only version.

25

Apps: For the web

Stepping outside Safari

There are several alternative browsers available in the App Store, but none that offers all the features and integration of Safari. Browse or search the Utilities and Productivity categories to see what's available, or try one of these:

Opera Mini

From the same people who make the brilliant desktop Opera browser, this iPhone version is impressively fast and features tabbed browsing, password tools and the quick access "Speed Dial" tool.

VanillaSurf

If you're not satisfied by Safari's Multiple Pages feature, this is the app to download for a very slick, full-screen tabbed browsing fix. It has tons of nice extra features, including a downloads manager, offline browsing and bookmark syncing. (For offline reading, also try the reader apps mentioned on p.206.)

Mercury Web Browser Pro

Another very handy Safari alternative. This one can be decorated with themes, features Firefox syncing tools and Dropbox integration, and has various handy finger gestures for swift browsing.

FREE Full Screen Private Browsing for iPhone & iPad

A no-frills affair that offers completely private browsing sessions: no history, no cache, no cookies etc. Though you can achieve the same thing by enabling the function for Safari (see p.215), you might find it easier to have a separate app available that always behaves in this way. Catchy app name too.

Site-specific apps

While some websites host optimized "mobile" sites or "webapps" for you to enjoy within Safari (often worth saving to your Home Screen as webclips, see p.74), many popular websites also dish up fully fledged apps. The advantage of such apps is that you're likely to have a far more polished and speedy interface at your disposal; plus, in many instances, some features are made available offline that would otherwise be unusable in a browser. They are generally free, so it's worth checking to see what your favourite sites offer in the App Store. Here are a few that are most definitely worth installing instead of using the regular websites:

Facebook

This is by far the best way to use Facebook on the iPhone, largely because it links in with the various notification features of the device, meaning that you always know what new activity is going on.

➔ **TIP** Under the Facebook app's Friends tab you'll find a useful button for syncing your iPhone's Contacts list with your Facebook Friends list.

LinkedIn

Another well-put-together social networking app that's worth installing if you regularly use this popular professional networking service.

eBay

A great little app that gives you access to all your buying and selling tools. If you sell on eBay regularly, you'll probably find the built-in selling tools easier to use than those on the regular desktop site, especially given that the iPhone's camera is so well integrated with the app. And if you're buying, you can get an immediate notification if your bid has been beaten.

➔ **TIP** Also, check out the Fat Fingers app that searches for commonly misspelled items on eBay to help you hone in on items that literate searchers miss out on.

Amazon Mobile

A well-managed app for shopping, reading reviews and making purchases from Amazon. Unfortunately, it doesn't allow you to purchase Kindle titles from within the app, though you can add them to your Wish List.

26
Staying safe online

Avoiding hacks, scams & theft

The Internet may be the greatest wonder of the modern world, but it does come with certain downsides. Increased access to information is great, but not if the information other people are accessing is yours – and it's private. Of course, there are threats to privacy and security in the real world, but on the Internet things are different, not least because your private data might be compromised without you even knowing about it. Although Apple products are generally very secure, it's worth remembering a few golden rules. If you want to keep your iPhone, your files and your private data safe, read the next couple of pages and make sure that you tick all the boxes.

• **Keep your system up to date** Many security breaches involve a programmer taking advantage of a security flaw in your software. So it's critical to keep your iPhone's software up to speed with the latest firmware updates (see p.259).

• **Don't run dodgy software** Apple work hard to ensure that all the software they make available in the App Store is "clean" of viruses and other dangers. However, if you're offered any alternative routes to getting software onto your iPhone, such as jailbreaking, steer clear – it's not worth the risk.

• **Don't respond to spam messages** Those "get paid to surf", "stock tips", "recruit new members", "clear your credit rating" and various other network-marketing schemes are always too good to be true.

> ➡ **TIP** Also beware of "phishermen" trying to snag your bank details and passwords.

• **Be careful of "adult" sites** It's often said that the majority of online scams involve porn sites – the scammers believing, probably correctly, that the victims will be too embarrassed to report the problem. If you do ever use an adult site, never pass over credit card details unless you're prepared to get stung. And, whatever you do, don't download any software they offer.

> ➡ **TIP** The iPhone's Restrictions features can be employed if you want to stop your kids accessing Safari on an iPhone. For the full story, see p.165.

Security on the road

Next, here are a few golden rules to help keep your iPhone's data safe and secure away from home, and particularly when connected to public Wi-Fi networks.

• **Always use a screen lock** Always use a password-protected screen lock so that, should your phone be lost or stolen, its data will be inaccessible. Look in Settings > General > Passcode Lock. Also turn on the Erase Data function so that all your iPhone's content is wiped after ten failed passcode attempts.

• Avoid using banking sites out and about If you use Wi-Fi hotspots, avoid accessing your bank accounts and any other sensitive material, as public networks are notorious for so-called "snoop" or "sidejacking" attacks, whereby data is intercepted by other machines using the same network. Where possible limit your activities to browsing that isn't security-sensitive. And always try to shield your keystrokes when entering passwords, just as you would at an ATM.

Find My iPhone

The "Find My iPhone" feature can be used remotely to track the location of your iPhone, to add a four-digit pass-code screen lock and even to wipe all its con-tent remotely – all very useful should your phone go astray. Go to Settings > iCloud to set up your phone so that it can be tracked.

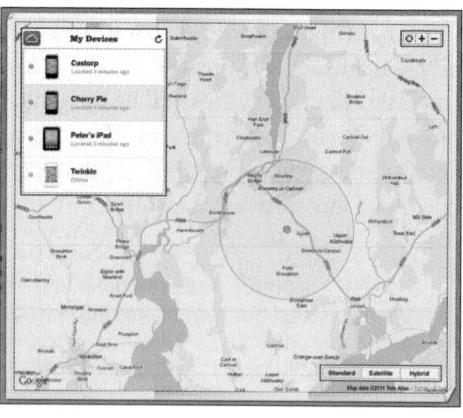

Assuming the iPhone is then connected to either a Wi-Fi or cellular network, you should be able to get a fix on it by logging onto the iCloud website (icloud.com).

> **→ TIP** There is also a Find My iPhone app that can be installed on any Apple iOS device, meaning you can track your iPhone from another iPhone, iPad or iPod Touch, as well as from the web.

Navigation

27

Maps

Search & directions

The Maps app on the iPhone (tap the icon on the Home Screen to get started) takes you into the world of Google Maps, where you can quickly find locations, get directions and view satellite photos. You can zoom and scroll around the maps in the same way you would with webpages in Safari, the difference being that, because you're using your fingertips to drag and pinch, the whole experience feels much more natural.

Searching for yourself

All recent iPhones can accurately determine your current location by using a combination of data from their GPS (Global Positioning System) chips, a connected cellular network and also from a connected Wi-Fi network.

To pinpoint your location, open Maps and tap the ◤ button at the bottom of the screen.

➔ **TIP** To zoom in, either double-tap one finger or spread two apart; to zoom out again, either pinch or tap once with two digits.

Which way am I facing?

To find out, hold the iPhone flat and tap the ◢ button for a second time; the app switches mode, allowing you to view the map in relation to the way you are facing. The tighter the angle coming out of your blue location marker, the more accurate the direction reading.

Searching for a location

Tap the Search box and type a city, town, region or place of interest, or a ZIP or postcode. (As you type, the iPhone will try to predict the location based on previous map searches and address entries in your Contacts list.) You can also try to find a business in the area you're viewing by entering either the name of the business or something more general – such as "camera", "hotel", or "pub". Note, however, that the results, which are pulled from Google Local, won't be anything like comprehensive.

Location Services settings

Whenever the app you are currently using is hooked into Location Services (via either GPS, Wi-Fi or the cellular network) a purple ◢ icon appears in the status bar at the top of the iPhone's screen. Look within Settings > Location Services to turn your iPhone's location functionality on and off for either specific apps or everything. This can be a useful means of saving battery power. You can also see here which apps have used the iPhone's Location Services in the last 24 hours as they display a purple ◢ icon.

→ **TIP** Where a search term that yielded multiple results has been used as an end point, tap the ≡ icon to quickly switch between the different pins and, in turn, refresh the route.

If a multitude of pins appears, tap on each of the pins in turn to see their name. Tap a pin's blue ◉ icon for further options, such as adding the location as a bookmark or contact address, emailing a link to the location or getting directions to or from that location.

→ **TIP** In some areas you will also see a little orange ⊕ icon on a pin's popover panel. Tap this to scroll around in Google's Street View. When you're finished, tap the inset map preview to finish.

Dropping pins

You can drop a pin manually at any time by tapping the page-curl button, bottom-right, and then Drop Pin. This can be a handy way in which to keep your bearings when sliding around a map. To change the position of a pin, tap and hold it and then drag. To delete the pin, add it as a bookmark, add it to a contact, email the location or get directions, tap the pin and then the ◉ icon. Your lists of bookmarks and recently viewed locations can be viewed by tapping the ⊞ icon in the right of the search field.

→ **TIP** The page-curl button also gives you access to a useful link for printing your current map view to any AirPrint-enabled printers (see p.246) on the same network as your phone.

Satellite and Terrain views

Tapping the page-curl button reveals options to view Satellite images (they're not live, unfortunately… maybe one day), a Hybrid view that adds roads and labels from the standard view to a satellite image, and also a List view, which can be a handy way to quickly take in all the pins that were generated by a search.

Directions

To view directions from one location to another, tap Directions at the bottom of the screen and enter start and end points, either by typing search terms or by tapping ⊞ to browse for bookmarks, addresses from your Contacts or recently viewed items. When you're ready, tap Route. Use the three icons at the top to choose between directions for driving, walking or taking the bus.

Once a journey is displaying, you can go through it one step at a time by tapping Start and using the arrow buttons to jump forward and back one stage. Alternatively, tap the page-curl button and then List to view all the stages as a series of text instructions. Tap any entry in the list to see a map of that part of the journey.

> → **TIP** For your return journey, reverse the directions given by your iPhone by tapping the ⇅ button between the start and end point fields.

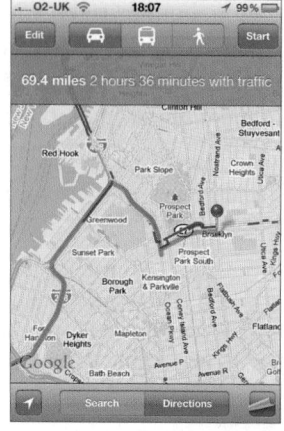

Traffic conditions

In areas where the service is available, your route will display colour-coded information about traffic conditions. The approximate driving time at the top of the iPhone screen will change to take account of the expected traffic and the roads will change colour:

- **Grey** No data currently available.

- **Red** Traffic moving at less than 25 miles per hour.

- **Yellow** Traffic moving at 25–50 miles per hour.

- **Green** Traffic moving at more than 50 miles per hour.

If you don't see any change in colour, you may need to zoom out a little. This action will also automatically refresh the traffic speed data. To hide the traffic information (perhaps you like surprises), tap the map curl and toggle the Traffic button off.

Searching for other people

Putting aside for one moment issues of personal space and privacy, devices such as the iPhone are now making it possible for you to geolocate your friends and family as and when you need to. Great for trying to meet up with people on a busy street; not so great if you told your other half that you were working late, when you're actually in a bar having, erm, another half!

Find my Friends

Apple's service requires you to download the app from the App Store and then invite friends to either "follow" you or see your location temporarily. You can also choose to make your location private as and when you need to.

Google Latitude

Google have an almost identical service in operation, also with an app available in the App Store.

28

Apps: Maps & travel

The best navigation tools

The App Store has its own categories dedicated to Navigation and Travel, where you'll find everything from plane and ship location tools (Plane Finder and Ship Finder) to multifunctional compass applications (Harbor Compass Pro; a great alternative to the iPhone's built-in Compass app). Have a browse to see what takes your fancy. Here are a few of our favourites:

Maps

Search "maps" in the App Store to unearth some interesting cartographic relics (Historic Earth) as well as some very impressive reference tools (National Geographic World Atlas). Also…

TomTom
These GPS navigation apps might be pricey, but they are still cheaper than buying the equivalent TomTom hardware.

iOS Maps

Beautifully rendered UK country-wide mapping, courtesy of Ordnance Survey.

Google Earth

An essential free app that gives you unfettered access to the globe. When you swipe with two fingers, or alter the tilt of your phone, you adjust the pitch of your view, which allows you to achieve some amazing views of mountainous regions. It can feel quite clunky at times, especially over a slow internet connection, as the higher resolution views can take a while to load, but it is worth being patient.

Transit Maps

Use this app to locate, download and store transit maps from around the world that you can then use offline. It's also worth looking into alternative city-specific transit apps, as many will give you the latest service information, assuming you have a web connection (worth checking before you head down the escalators to the subway station). For the London Underground the best app is Tube Deluxe.

London A-Z

Brilliant sets of London street maps (there are several editions to choose from, depending what you need). All the mapping is stored offline so you don't even need a web connection to use it. There are equivalent offline maps available in the App Store for most major cities – essential if you know you are going abroad and don't want to rack up roaming charges by using Google Maps.

General travel

The main drawback of travelling with an iPhone is the issue of roaming charges, so if you do download destination guides before you depart, make sure they include offline content and maps (as is the case with those published by Rough Guides), so that you can use them with data roaming turned off (Settings > General > Network).

Wikihood
Offering location-based access to Wikipedia, this app is great for finding out about buildings around you. It's not so good overseas though, as it requires an internet connection.

Rough Guides Travel Survival Kit
This app covers everything from how to avoid getting mugged to what to do if you come face to face with a bear. It also has a useful section for storing visa and passport numbers, as well as quick-dial links to embassies and your next of kin.

Tripit
Useful app that aggregates all your trip information (flights, transfers, hotel bookings etc) into one place.

Skyscanner
Great app for finding the cheapest flights. It searches across nearly all the airlines and resellers and serves up the results in an easily digestible form.

British Airways
Most airlines now have their own app. The BA one lets you manage your flights and air miles and also lets you use your iPhone as a paperless boarding pass.

App
essentials

29

Apps: Productivity

Tools for everything

Like most modern phones, the iPhone comes with a small menagerie of extra tools such as alarms, a calculator, a weather app and a notes tool. But thanks to the App Store, the iPhone apps that come built in are really just the tip of a hulking great iceberg...

Apps that keep you organized

Clock
To check the time on your iPhone, look to the digital clock which is almost always present at the top of the screen. The Clock app, meanwhile, offers a World Clock (tap **+** or Edit to add locations) and a self-explanatory Stopwatch, Alarm and Timer.

→ **TIP** Use Siri (see p.53) to quickly set the timer, using a phrase such as: "Set timer for fifteen minutes..."

> **→ TIP** Even if you have your iPhone in silent mode, the alarm will still sound when it is due to go off – which is one less thing to worry about when setting your alarm at bedtime, but could, potentially, cause embarrassment in the cinema or a meeting.

Reminders

The best thing about the iPhone's built-in Reminders app is that it can be synced with iCloud (turn it on within Settings > iCloud) and also iCal and Outlook, so that you can make changes to your lists on multiple devices and your computer. You can also create location-based reminders to pop up on your iPhone, so, for example, your phone may remind you to buy something when you enter the supermarket.

> **→ TIP** Try setting reminders using Siri – it's a lot faster than using the keyboard.

GeeTasks

A slick little app for Google Task syncing, complete with a Home Screen badge icon for uncompleted tasks.

> **→ TIP** You can set an alert to remind you of an impending event either as it happens or a certain number of minutes, hours or days beforehand. If you'd like to have these alerts presented visually, instead of via a sound, tap Settings > Sounds and turn off Calendar Alerts.

Remember The Milk

This app is arguably the best task-organizing sync service out there right now. You do have to sign up for a fee-paying Pro account to make use of the iPhone app (though, like Apple's Reminders, it does support Siri voice interaction).

Note-taking apps

Notes

Notes is Apple's simple application for jotting down thoughts on the go. Tap ✚ to bring up a blank page and you're ready to type. When you're finished, tap Done or use the ✉ icon to email or print (see p.246) the note. Look within Settings > iCloud to turn on syncing, and Settings > Notes to choose a font.

Awesome Note

A really stylish notes and to-dos tool with colour-coded folders and lists.

Evernote

Evernote is among the best note-taking tools and services available. The iPhone app can create notes from text, images and audio and these notes are then synced between the Evernote server and whatever desktop or mobile versions of Evernote you are running.

Voice Memos

For voice notes, don't forget that the iPhone comes with an excellent memo recorder built in. Recorded memos can be sent via email or MMS from the iPhone, and also synced back automatically to a special Voice Memos playlist in iTunes when you connect and sync, either wirelessly or via a cable.

➜ **TIP** You can use Calendar instead of a notes app to jot down thoughts on the move. Just add an event and use the Notes field. You can set a date and time when you think the jottings will be useful, and even set an alert.

Writing apps

Pages

Apple's word process-ing app is very slick, but pricey compared to others. It does integrate nicely with iCloud storage however (meaning that you can pickup where you left off on another device), and is a universal app, so it will also run on the iPad.

Doc²

The formatting tools of this word processor app (pictured) are really impressive, offering everything from bold, italic and underline to bullets, indents and table construction. Its best feature is the fact that it works with Google Docs to help you share your files online.

WordPress

For the WordPress bloggers out there, the iPhone app is a must: both well designed and easy to use. As well as creating and editing your posts, you can also use the tool to moderate comments and add audio and video.

WriteRoom

This app offers a distraction-free writing environment where you can get on with your writing without toolbars and formatting worries. It works in a white-on-black mode that makes it easy on the eye for extended periods of writing.

Pocket CV

This app makes sure you always know where your résumé is, and more importantly, can keep it up to date. Its layout features are impressive, meaning you can email off a fully fledged PDF CV at the drop of a hat from anywhere.

Money apps

Stocks

The built-in Stocks app lets you view current values for any listed company. Well, not quite current – the prices are refreshed whenever you open the application, but are typically still about 20 minutes out of date.

→ **TIP** The companies that you add to the Stocks app also appear as a live ticker in Notification Center (see p.52). To disable this ticker, visit Settings > Notifications > Stock Widget.

XE Currency

XE Currency is the only currency converter tool you need. It really comes into its own when you're abroad and need to quickly check whether or not something is a bargain.

Money Smart Lite

One of many budgeting apps that help you keep an eye on your spending. You enter your budget and then document all your spending within different categories. You can also create graphs to see your spending patterns over the week, and manage savings and bills as well.

Weather apps

Weather
The iPhone's built-in Weather application does pretty much what you would expect. You can save favourite locations and view current conditions along with a six-day forecast. For a quick update of the weather, look to Notification Center (see p.52).

Met Office Weather application
For the UK, this Met Office alternative is a must, largely because it includes radar and satellite imagery. Also, the five-day forecast tends to be more accurate than the one on the Apple app (which pulls its data from Yahoo!).

Fahrenheit & Celsius
These two separate apps keep you up to speed with the current temperature and handily display as a badge icon.

Number apps

Calculator
Paying no small tribute to the classic designs of German functionalist Dieter Rams, the look of the iPhone's Calculator app can be traced back to a calculator that Braun (for whom Rams worked) built for Apple in the 1980s. When in landscape, it transforms into scientific mode.

PCalc Lite
Though you can arguably get everything you need from the built-in Calculator app, PCalc Lite is free, looks nice, has scientific functions and can also handle unit conversions.

→ **TIP** If you have an iPhone 4S, also try Siri (see p.53) for voice-activated unit conversions.

Numbers

This iWork application is Apple's answer to Excel. Though not as sophisticated as the Microsoft desktop program in terms of advanced spreadsheet features, it is nicely put together and does an impressive job of creating charts and graphs. The templates are excellent and you can move between dry-looking data and something eye-popping with little effort. The Share & Print function allows you to send files via the iTunes file sharing mechanism (see p.249) as either a PDF, Numbers or Excel document. You can also attach documents to emails straight from the app.

→ **TIP** Documents can additionally be uploaded from any of the iWork suite apps to the iCloud website (icloud.com). In any of the apps, view the document gallery, tap Edit, highlight a selection, and then tap the ⬜ icon to see the option of posting to iCloud.

Quick Graph

A free app that can be used to produce both 2D and 3D graphs from complex formulas and equations.

Sheet[2]

Crafts fairly impressive spreadsheets and can both edit and create Microsoft Excel docs. It also works with Google Docs for sharing your files online.

→ **TIP** If you like the sound of both Doc[2] and Sheet[2], check out the Office[2] app, which combines both at a cheaper price.

Presentation apps

Keynote

The third member of Apple's iWork toolkit is Keynote: a stylish equivalent of PowerPoint from Microsoft Office. What it lacks in features it makes up for in ease of use and elegant templates. It can open and edit PowerPoint documents, and you can also save Keynote projects to iCloud and then download them to a desktop machine as either Keynote, PDF or PowerPoint documents.

> **→ TIP** You can connect your iPhone to a projector and use Keynote to run your presentation as well as create it, using an iPhone Dock Connector adapter (see p.185).

Keynote Remote

This Apple app lets you control a presentation you are running from Keynote on either another device or a Mac on the same Wi-Fi network as your iPhone.

MyPoint PowerPoint Remote

As above, but this time the app is for controlling PowerPoint presentations playing on either a Mac or a PC.

ProPrompter

With this app you can turn the iPhone into a script teleprompter and, with the same app installed on another device (either iPad, iPhone or iPod Touch), you can control the pace of the script on the first device via a Bluetooth connection.

Artistic apps

Brushes – iPhone Edition

Brushes is a beautifully crafted app that artists of any level will have a lot of fun with. It supports image layering, different brush textures and weights, and the ability to control the transparency and weight of your strokes based on the speed with which you move your finger. Also featured is a very cool "replay" function that lets you watch the progress of your creation as a movie once you're done. You can also import images from the Photos app to incorporate into your artwork.

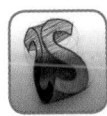

SketchBook Mobile

Like Brushes, this app supports layers and is really intuitive to use. Overall, it's probably more suited to professionals, offering loads of tools, brushes and textures, and an incredible zoom feature, which allows you to get right in there and add detail.

123 Color HD

There are loads of colouring apps aimed specifically at kids. This one is a real treat – it has songs, voiceovers and multiple languages built in (so it can be used for basic language teaching too). When they're done, pictures can either be saved to Photos or emailed to granny and grandad.

Spawn Illuminati HD

This one is for all the hippies out there – by tapping and swiping, you control the creation of multicoloured psychedelic patterns on the iPhone screen. There's even a public Flickr gallery (flickr.com/groups/spawn) where you can post your creations for the world to see.

Printing from apps

The iPhone supports Apple's over-the-airwaves print technology AirPrint, which allows you to print directly from the iPhone to your printer over a Wi-Fi network. Unfortunately, you have to own one of several pricey HP printers to make it work. For a full list of compatible machines, visit apple.com/uk/iphone/features/airprint.html. Once you have the new printer set up, AirPrint printing is found by tapping the  icon (in many apps), selecting the number of copies you want, and then tapping Print.

There are, however, many apps that offer printing functionality without the need to buy a fancy new printer.

Printopia

If you have a Mac computer with a printer connected to it, this is by far your best option. Once this Mac application is installed on your computer, your iPhone will recognize your printer in the same way it would recognize an AirPrint printer within the  options menu. What's more, it also allows you to send the print job to your Mac as a JPG or PDF, or, if you use Dropbox (see p.248), directly to your Mac's Dropbox folder. Printopia can be downloaded from ecamm.com/mac/printopia.

PrinterShare Mobile

This app lets you print from the iPhone – over the airwaves – photos, contacts' details, webpages, and the contents of the iPhone's clipboard (i.e. anything clipped using the Copy command). For the full story, visit PrinterShare.com.

Fax, Print & Share Pro

Does exactly what it says on the tin, working with both network printers and "shared" printers connected to a desktop machine. The fax feature is a useful bonus. The setup process can be a little fiddly, but it's worth the effort.

Utilities & file storage

Jotnot Scanner Pro

A useful scanning tool. Use the iPhone's camera to take a photo of a document and then store or share the resulting file. Shadows are automatically removed.

Air Display

Turns your iPhone into an extra screen for your Mac. Use it, for example, to display a small-window application such as iChat while you use your Mac screen for other purposes.

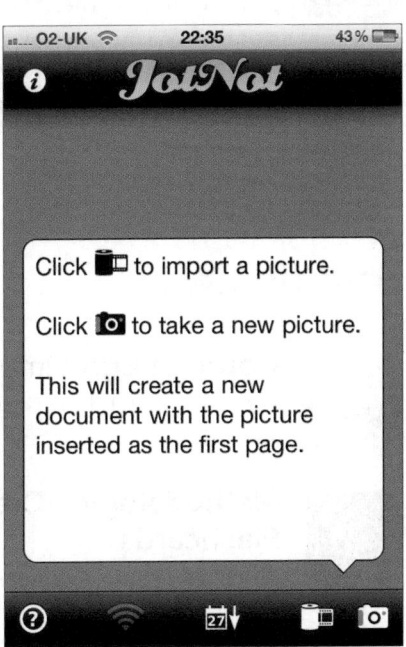

Dropbox

Dropbox is probably the best online file-storage system, making it a cinch to sync files and folders among multiple computers. Best of all, it's free – and the iPhone app allows you to view the contents of your Dropbox folder from your phone.

Air Sharing

This excellent app gives onboard file storage, Wi-Fi file sharing with desktop machines and access to networked printers. The universal version also runs on the iPad.

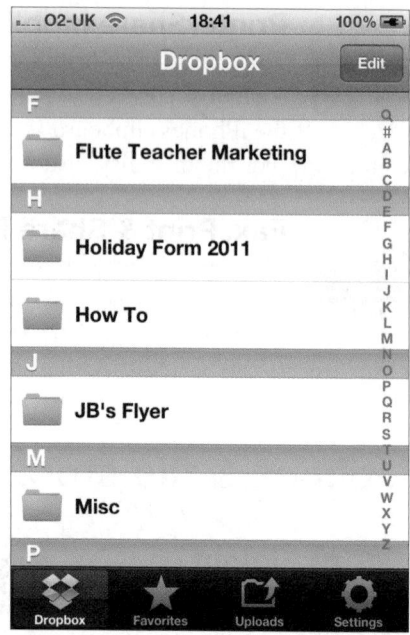

And finally...

Gourmet Egg Timer

For the perfect boiled egg, this app takes into account altitude, egg size and how soft you like your yolk.

Methodology – Creative Inspiration Flashcards

Modelled on the "Oblique Strategies" deck of cards created and originally published back in 1975 by Brian Eno and

File sharing with apps via iTunes

Many apps, including the three members of the iWork suite, can utilize the File Sharing feature of iTunes as a means of transferring files back and forth between the iPhone and your computer. If an app supports the feature, it will appear within the File Sharing panel

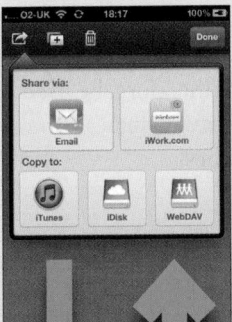

in the lower area of the Apps tab in iTunes when your iPhone is connected and highlighted in the iTunes sidebar. From there you can drag files in (or use the Add button to browse for files) and drag files out (or use the Save to button to browse for a location to save to).

If you can't see your iWorks files in iTunes, you will first have to go back into the respective app, head to the documents gallery, tap Edit, make a selection and then tap ☛ to get the option to move your document to the iTunes File Sharing panel.

Similarly, to pull documents into the iWorks apps, go to the app's document gallery, tap the **+** icon and choose the iTunes option.

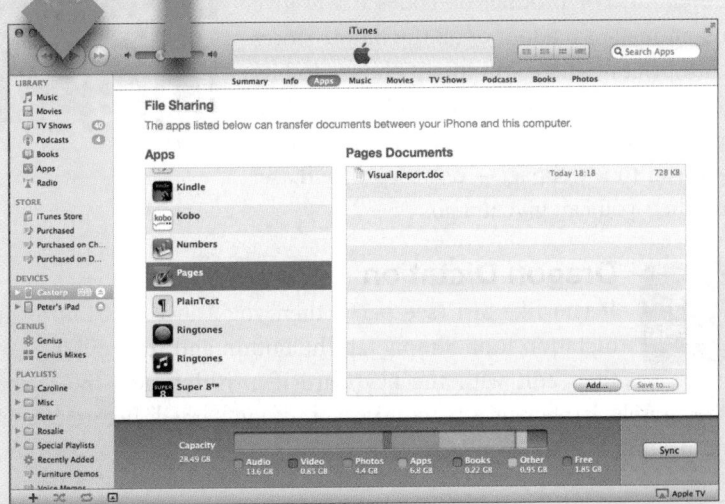

If you would rather share files between your iPhone and computer over Wi-Fi without iTunes, use an app such as Air Sharing, and for sharing via the web, Dropbox or iCloud.

Peter Schmidt, this app presents a set of cards that, in their own words, "prompt you to think differently". They can be really effective (and at other times simply amusing) for those moments when inspiration just won't come.

FreeSpirit

Genius. Your iPhone has a straight edge… your iPhone has an accelerometer… your iPhone can, therefore, be used as a spirit level. There are loads of apps that can do it, so you may as well base your choice on which icon you like best.

RedLaser

An essential app for reading bar-codes and QR codes (such as the one shown here) when you're out and about. For product barcodes it will give you an instant readout of all the online stores (with prices) where you might choose to buy the item. QR codes, meanwhile, are often used on posters or in magazines to quickly take you to a webpage – download RedLaser and then you can give it a go.

Dragon Dictation

If you like Siri (see p.53), then you'll love this impressive dictation tool. Simply tap the button and start talking, and then edit with the keyboard if anything gets transcribed inaccurately. It's a great way to quickly draft long emails before copying and pasting into the Mail app.

30

Apps:
Games

Killing time on the iPhone

Both iPhones and iPods before them have always come with a smattering of games built in. But with the arrival of the App Store, gaming on Apple hardware became much more than just light distraction. Playing games on iPhones, iPods and iPads has become so popular that whole new brands have grown up around it – the  must-have Angry Birds being an obvious example. Today, over fifty percent of the top-ranking apps in the Store at any one time are games. And it's not just traditional action-packed videogames – blood-splattered zombies and high-octane racing – that are being consumed in vast quantities (though if that's your vice, you won't be disappointed). In this chapter, we'll look briefly at everything from fiendishly challenging puzzles to table tennis.

Puzzles & strategy

Touch Physics 2

Using gravity and various charming crayon-sketched objects, you have to defeat devious level after devious level. Once you get started, don't expect to talk to anyone for a good few days.

Chess Free

Dozens of chess apps are available for the iPhone. This one is free, looks nice and does everything you need it to.

Shantae: Risky's Revenge

Familiar to some, perhaps, from the earlier incarnation of Shantae on the Nintendo Game Boy, this is a great first step into the world of iPhone platformers.

Cut The Rope

A beautifully rendered puzzle game that has you feeding candy to a devilish little green critter. A great example of a game that is perfectly suited to finger play on a small screen.

Apple Game Center

Apple Game Center is a social-networking tool that many app developers build into their games to enable global leaderboards and multiplayer gameplay with your friends. There is also an auto-match feature that can help you find new people to play a given game with. The Game Center app can be found pre-loaded on your iPhone; it gives you access to all your scoreboards and also shortcuts to the actual game apps (tap through and hit the Play buttons). Find out more at apple.com/game-center.

Action & adventure

Modern Combat 3: Fallen Nation

The quality of the graphics is jaw-dropping; this really is among the best first-person shooters currently available for the iPhone.

Galaxy on Fire 2 HD

The fact that this science-fiction epic only runs on the iPhone 4S illustrates just how processor-heavy the gaming environment is. You are taken through a seemingly boundless galaxy of strange worlds and mammoth space stations. Well worth checking out.

> **→ TIP** The best place to go for the latest iPhone games reviews and news is toucharcade.com.

Racing

Drag Racing Free

Well-respected racing game with integrated Game Center duelling. Head into Settings and turn on vibrations to add a new dimension to the gameplay.

Nano Rally

This fun little racing game sees your tiny vehicle going head to head with other minuscule motors. If you ever played the classic Micro Machines, and like interesting driving environments, you'll love this.

Retro gaming & arcade

Atari's Greatest Hits
This app gives you access to a large library of old Atari games from both the arcade and console vaults.

Sports

FIFA 12
It's hard to fathom how they managed to get so much detail and sophisticated gameplay into this iPhone game. A must for all soccer fans and devotees of the console versions.

World Cup Table Tennis
Highly addictive, this iPhone version of ping pong really sucks you in. It hooks up with Game Center, so that you can battle with opponents from around the world to win the championship.

→ **TIP** If you have an Apple TV and an iPhone 4S, try using screen mirroring and AirPlay (see p.186) to play your games through a connected HD TV screen or projector. Though latency can be an issue, it is worth experimenting.

Extras

Maintenance

Troubleshooting & battery tips

The iPhone is a tiny computer and, just like its full-sized cousins, it will occasionally crash or become unresponsive. Far less common, and much more serious, is hardware failure, which will require you to send the phone away for servicing. This chapter gives advice for both situations, along with tips for maximizing battery life.

Crashes and software problems

Every now and again you should expect your iPhone to crash or generally behave in a strange way. This will usually be a problem with a specific application and the iPhone will simply throw you out of the app and back to the Home Screen. From there, simply tap your way back to where you were and start again. If the screen completely freezes, however, try the following steps:

• **Force-quit the current application** Hold down the Sleep/Wake button until the red slider appears, then immediately let go of that button and hold the Home button for five or six seconds.

• **Reboot** As with any other computer, turning an iPhone off and back on often solves software glitches. To turn the phone off, press and hold the Sleep/Wake button for a couple of seconds and then slide the red switch to confirm. Count to five, and then press and hold the Sleep/ Wake button again to reboot.

• **Empty the app switching tray** Double-tap the Home button and then tap and hold one of the items in the app switching tray (see p.71) until they all start to jiggle. Now use the ⊖ icons to remove the apps from the tray. You aren't deleting the apps from your phone, just their saved states. Doing this can really help with a sluggish phone as it frees up the device's memory.

• **Reset** If that doesn't do the trick, or you can't get your phone to turn off, try resetting your phone. This won't harm any media or data on the device. Press and hold the Sleep/Wake button and the Home button at the same time for around ten seconds. The phone may first display the regular shutdown screen and red confirm switch; ignore it, and keep holding the buttons, only letting go when the Apple logo appears.

> **→ TIP** If you are having problems with Wi-Fi connections, try tapping the Settings > General > Reset > Reset Network Settings button. You will then need to re-enter passwords for previously remembered networks.

• **Reset all settings** Still no joy? Resetting your iPhone's preferences could possibly help. All your current settings will be lost, but no data or media is deleted. Tap Settings > General > Reset > Reset All Settings.

• **Erase all content** If that doesn't work, you could try deleting all the media and data too, by tapping Settings > General > Reset > Erase All Content and Settings. Then connect the iPhone to your computer and restore your previously backed-up settings (see p.260) and copy all your media back onto the phone.

• **Restore** This will restore the iPhone's software either to the original factory settings or to the settings recorded in the most recent automatic backup (see p.260). In both cases, all data, settings and media are deleted from the phone. Connect the iPhone to your computer and, within the Summary tab of the phone's options pane in iTunes, click Restore and follow the prompts.

> ➔ **TIP** An iTunes restore will only work if your computer is connected to the Internet, as the device needs to be "verified" with the iTunes servers.

• **Software update** You should be prompted to update your iPhone's internal software, or operating system, automatically from time to time. But you can check for new versions at any time from the Settings > General > Software Update screen. This can also be done within iTunes by clicking the Check for Update button on the Summary tab when your iPhone is highlighted in the sidebar. If a new version is available, be sure to install it.

iPhone firmware

As well as the software operating system (iOS) mentioned above, the iPhone also runs network carrier firmware (carrier settings) that allows it to connect to a given supplier's phone and data networks. If you ever switch network providers and install a new SIM card, connect to iTunes and the carrier settings will be automatically updated. From time to time you might be offered an update to your carrier settings even when you haven't changed the SIM. This doesn't affect your own settings, media or data, and is definitely worth installing.

Backing up

It is important to create a backup of the data, settings and configuration of your iPhone, including your mail settings, messages, notes, call history, Favorites list, sound settings and other preferences. From a saved backup, you can restore your device to its previous state should it need to be reset to factory settings. A backup can also help you to set up a new iPhone with the same configuration as an older device as and when you upgrade to a newer model. There are two ways to set up automatic backups:

Backing up to iCloud

Look within Settings > iCloud > Storage & Backup and slide the iCloud Backup switch to the On position. Alternatively, connect to iTunes on your computer, highlight your iPhone in the sidebar and look for the option under the Summary tab.

To see more information about your iPhone's backup, and most importantly, how much of your iCloud storage space is being consumed by the backup, tap the Manage Storage button. Next, tap through to see exactly what's being backed up from your device. There is a separate listing for every app and you can choose to stop backing up specific apps if you wish (perhaps if an app is taking up a lot of space in iCloud and you are not that worried about losing its data).

> ➔ **TIP** Note the green strip at the bottom of the screen that displays how much of your iCloud quota is left. If you need more storage space, tap the Buy More Storage button.

Backing up to iTunes

To back up to iTunes on your computer, connect your iPhone and look under the Summary tab in iTunes for the option that reads "Back up to this computer". You can also set an encryption password so that anyone with access to your computer will not be able to restore their devices to your backup and overwrite it.

To view, and if necessary delete, an automatic iPhone backup in iTunes, open iTunes Preferences and click Devices. Of course, a backup stored in iTunes rather than iCloud is only as safe as your Mac or PC. Computers can die, get destroyed or be stolen, so get into the habit of backing up your computer to either an external hard drive or cloud-based computer backup service. For more on this subject, see *The Rough Guide to Cloud Computing*.

Automatic and manual backups

If you use the iCloud method, your backup will be automatically updated whenever your iPhone is plugged in, locked and connected to Wi-Fi. While connected to Wi-Fi you can also manually kick-start a backup via the button on the Settings > iCloud > Storage & Backup screen.

An iTunes backup will happen automatically whenever your iPhone is connected to iTunes via either a cable or your Wi-Fi network. To set one going manually, right-click your iPhone's entry on the iTunes sidebar and select the option to Back up.

The iPhone battery

There has been a fair amount of controversy, and no shortage of misinformation, about the iPhone's non-user-replaceable lithium-ion battery (see p.16). Like all lithium-ion batteries, the one inside the iPhone lasts for a certain amount of time before starting to lose its ability to hold a full charge. According to Apple, this reduction in capacity will happen after around 400 "charge cycles".

Despite various newspaper reports to the contrary, a charge cycle counts as one full running down and recharging of the battery. So if you use, say, 20 percent of the battery each day and then top it back up to full, the total effect will be one charge cycle every five days; in this situation the battery would, theoretically, only start to lose its ability to hold a full charge after quite a few years: 5 days x 400 charge cycles = 2000 days.

If, on the other hand, you drain your battery twice a day by watching movie files while commuting to and from work, you might see your battery deteriorate after just nine months or so. In this case, you'd get it replaced for free, as the iPhone would still be within warranty.

> → **TIP** To turn on or off the percentage charge indicator at the top of the screen look for the setting within Settings > General > Usage.

If the battery won't charge

If your iPhone won't charge up via your computer, it could be that the USB port you're connected to doesn't supply enough power or that your Mac or PC is going into standby mode during the charge. To make sure the phone is OK, try charging via the supplied power adapter. If this doesn't work either, it could be the cable; if you have an iPod cable lying around, try that instead, or borrow one from a friend. If you still can't get it to charge, send the phone for servicing (see p.265).

Tips for maximizing battery life

Below you'll find various techniques for minimizing the demands on your iPhone battery. Each one will help ensure that each charge lasts for as long as possible *and* that your battery's overall lifespan is maximized.

• **Keep it cool** Avoid leaving your iPhone in direct sunlight or anywhere hot. Apple state that the device works best at 0–35°C (32–95°F). As a general rule, try to keep it at room temperature.

• **Keep it updated** One of the things that software updates can help with is battery efficiency. So accept any that are offered.

• **Drain it** As with all lithium-ion batteries, it's a good idea to run your iPhone completely flat once a month and then fully charge it again.

• **Dim it** Screen brightness makes a big difference to battery life, so if you think you could live with less of it, turn down the slider within Settings > Brightness. Experiment with and without the Auto-Brightness option, which adjusts screen brightness according to ambient light levels.

• **Quit multitasking apps** Any apps that are doing their own thing in the background will be draining your battery, so if your battery is low, clear the app-switching tray (see p.71).

• **Turn off notifications** Head to Settings > Notifications and turn off notifications for any apps you really don't need to be hearing from.

• **Tweak your Fetch settings** Within Settings > Mail, Contacts, Calendars > Fetch New Data, turn Push to off and choose to Fetch Manually (i.e., when you open the Mail or Calender apps) or Hourly.

• **Turn off Location Services** As above, if you need to conserve energy when out and about, turning this off (Settings > Location Services) will really help to extend your charge.

• **Lock it** Press the Sleep/Wake button when you've finished a call to avoid wasting energy by accidentally tapping the screen in your pocket before the phone locks itself.

• **Turn off Wi-Fi, 3G & Bluetooth** These power-hungry connections are easily turned off when not in use. Use the switches under Settings > Wi-Fi, Settings > General > Network, and Settings > General > Bluetooth.

• **Junk the EQ in the Music app** Set your iPhone to use Flat EQ settings (under Settings > Music). This will knock out imported iTunes EQ settings, which can increase battery demands.

• **Stay lo-fi** High-bitrate music formats such as Apple Lossless may improve the sound quality (see p.151), but they also increase the power required for playback. If you sync your music from iTunes, look for the option under the Summary tab to automatically convert higher bitrate songs to 128 kbps.

Warranty, AppleCare+ and insurance

The iPhone comes with a one-year warranty that covers everything you'd expect (hardware failure and so on) and nothing that you wouldn't (accidents, loss, theft and unauthorized service). The only departure from the norm is that while most Apple mobile products come with a warranty that enables you to take the faulty device to any Apple dealer around the world, the iPhone warranty currently only covers the country in which you purchased it.

In addition, you can choose to extend your warranty by a further two years through the AppleCare+ scheme. The price at the time of writing is $99 (£61 for the AppleCare Protection Plan in the UK). This service includes battery cover and two accidental damage repairs (with a $49 surcharge) – which could be handy if you're prone to screen breakages.

AppleCare apple.com/support/products

As for insurance against accidental damage and theft, most network carriers in most countries do offer insurance, albeit for a fairly high fee. Alternatively, you could investigate what options are available via your home contents scheme. Many insurers offer away-from-home coverage for high-value items – though this can also be expensive.

When the battery dies

When your battery no longer holds enough charge to fulfil its function, you'll need to replace it. The official solution is to send the phone to Apple, who, if your phone is no longer within its warranty period, will charge you around $79/£55 plus shipping. You'll get the phone back after three working days, but if you can't wait that long, you could hire a replacement for $29/£20.

Apple apple.com/batteries/replacements.html

The unofficial solution is to try to find a less expensive battery from a third-party service, along with a fitting service or DIY instructions. At the time of writing, this isn't an option, but it seems likely that the companies that already offer iPod battery replacements will soon expand their range. These include:

iPod Battery ipodbattery.com
iPod Juice ipodjuice.com

iPhone repairs

If the troubleshooting advice in this chapter hasn't worked, try going online and searching for help in the sites and forums listed on p.268. If that doesn't clear things up, it could be that you'll need to have your phone repaired by Apple. To do this, you could take it to an Apple retail store (see p.30), though you may have to make an appointment.

Alternatively, visit the following website and fill out a service request form. Apple will send you an addressed box in which you can return your phone to them. It will arrive back by post.

Apple Self-Solve selfsolve.apple.com

Either way, you're advised to remove your SIM card before sending it off (see p.37) and you can choose to hire a replacement iPhone for the period of repair ($29 in the US, £20 in the UK). But make sure you return it on time, or you'll be charged a fee, or the full price of the phone, depending on the length of the delay.

iPhone screen replacement

Arguably the most common physical ailment to befall an iPhone is a cracked or shattered screen. In most instances, the screen will continue to function normally and the device can be used, but it is very annoying so you will want to get it fixed. There is a lot of misinformation online, so briefly, these are your options, from expensive to cheap:

• **Get Apple to repair it** This repair is not covered by the standard warranty, and doesn't come cheap. If you have AppleCare+, you will be covered (see box on p.264).

• **Get someone else to repair it** Various websites (such as iresq.com) offer cheaper repair services, though paying them to do the work will more than likely void any remaining warranty.

• **Repair it yourself** If you are feeling brave, have a steady hand and don't mind voiding your warranty, you could opt for one of the DIY kits to be found online. Amazon is a good place to start, but do make sure that you read the reviews of the kit before you buy, and make sure you get the right one for your model of iPhone.

→ **TIP** Another excellent source of screen repair knowledge, kits and tools is ifixit.com.

Diagnostic codes

Many phones respond to diagnostic codes – special numbers that, when dialled, reveal information about your account, network or handset on the screen. Assembled in part from a post at The Unofficial Apple Weblog (tuaw.com, essential reading for all Mac, iPod and iPhone users), the following list includes many such codes for the iPhone. Some are specific to a given carrier, or the iPhone, while others will work on any phone.

Note that these are American codes. European iPhone users may find that only a few of them work on their handsets.

*#06# Displays the IMEI – the handset's unique identification code.

*225# Call Displays current monthly balance.

*646# Call Displays remaining monthly minutes.

*777# Call Displays account balance for a prepaid iPhone.

*#61# Cal Displays the total number of calls forwarded to voicemail when the phone was unanswered.

*#62# Call Displays the total number of calls forwarded to voicemail when the phone had no service.

*#67# Call Displays the total number of calls forwarded to voicemail when the phone was engaged.

*#21# Call Displays various Call Forwarding settings.

*#30# Call Displays whether or not your phone is set to display the numbers of incoming callers.

*#43# Call Displays whether call waiting is enabled.

*#33# Call Displays whether call barring is enabled.

3001#12345# Call Activates Field Test mode, which reveals loads of hidden iPhone and network settings and data. Don't mess with these unless you know exactly what you're doing.

Useful web sites

If you have an ailing iPhone, or if you want the latest tips, tap Safari and drop in to one of these support sites…

Apple Support apple.com/support/iphone
iPhone Atlas iphoneatlas.com
MacFixit macfixit.com

… or pose a question to one of the iPhone junkies who spend their waking hours on forums such as:

Apple Forums discussions.apple.com
Everything iPhone everythingicafe.com/forum
iLounge forums forums.ilounge.com

If you're the kind of person who likes to take things apart, check out the "Cracking Open the Apple iPhone" article at TechRepublic, and follow up with one of the numerous "iPhone disassembly" or "iPhone mod" posts on YouTube and Flickr. Trying any such thing at home will, of course, thoroughly void your warranty.

TechRepublic techrepublic.com
Flickr flickr.com
YouTube youtube.com

> ➜ **TIP** If you have a suggestion about how the iPhone could be improved, then tell Apple at: apple.com/feedback/iphone.html

Useful apps

Speedtest.net Mobile Speed Test
Use this app to measure the speed of your iPhone's current internet connection; it works for measuring both Wi-Fi and carrier network speeds.

Battery
There are loads of apps that can display in detail how much time you have left to do different activities based on your current charge. This is one of the more handsome ones, and it also has the added feature of an alarm that sounds when your iPhone reaches 100% while being recharged.

System Manager
Nice tool for looking to see how much of your system's resources are being used by running applications. It also gives you battery diagnostic readouts.

32

Accessories

Plug and play

There are scores of iPhone accessories available, from FM transmitters to travel speakers. The following pages show some of the most useful and desirable add-ons, but new ones come out all the time, so keep an eye on the iPhone sites listed on p.246. When it comes to purchasing, some accessories can be bought on the high street, but for the best selection and prices look online (see p.278). Compare the offerings of the Apple Store, Amazon, eBay and others, or go straight to the manufacturers, some of which sell direct. Before buying any accessory, make sure it is definitely compatible with your model of iPhone.

Cases and screen protectors

The iPhone is more resistant to scratches and other day-to-day wear than many other smartphones. But it's not invulnerable, and screen cracks are not uncommon, so you might want to protect yours with some kind of case and possibly a screen protector such as the iPhone InvisiShield ($15/£10 for a pack of two).

As for cases, you can either go for a minimal design, such as Apple's own Bumpers (pictured) for iPhone 4 and iPhone 4S; these protect all the corners, but leave the back of the phone exposed, and are great if you don't want too much bulk in your pocket.

If you are after a case that is more substantial (perhaps something fat and rubbery that might bounce when you

drop it), then take a look at the Incipio BOMBPROOF (pictured). Or alternatively something more designed (in leather that folds out like a wallet perhaps). There are literally hundreds of case designs out there, so do some research and find one that suits you and your lifestyle.

If you want to use your iPhone when jogging or in the gym, look for something with an armband (such as Belkin's Sports Armband) or a belt clip.

Wired headsets

The stereo headset that comes bundled with the iPhone is functional enough, but not ideal. Even if the audio quality were better, it still wouldn't be especially comfortable to wear. Many people can't even get the earphones to stay in place.

Thankfully, there are numerous alternatives out there which are comfortable, "noise isolating" (that is, they snugly fill your ear thereby blocking out ambient noise) and have a minijack plug svelte enough to be used with most iPhone cases. The best budget option is Rivet's Stereo Headset, which offers superior sound, lanyard-style cords, a decent mic and a choice of interchange-able earbuds.

If you're serious about sound quality, you might prefer something from V-MODA or Shure (pictured), both of whom create very high quality products that will not disappoint.

Bluetooth headsets

Bluetooth headsets let you receive calls without getting out your phone or wearing a wired headset. They sit in or around your ear, respond to voice commands and communicate with your phone via radio waves. The main problem is that they can make you appear to be talking to yourself – so expect some strange looks in the street.

Apple's own offering was released at the same time as the first iPhone but quietly discontinued in 2009 – presumably because of low sales. There are many other brands still available, though. One slick-looking but pricey option is the Jawbone ERA. It's comfortable and boasts some clever built-in software that adjusts the volume to account for the noise of your environment. You can also keep an eye on its battery level via a special icon that pops up on the iPhone screen.

For those on a tighter budget, most other Bluetooth headsets should work fine. They start at around $50/£25, though the very cheapest ones tend to suffer from poor battery life.

Keyboards

If you have an iPhone 4S, you will be able to connect to a Bluetooth keyboard for typing, which can be incredibly useful if you want to use an iPhone as your primary email device. Apple make an excellent Bluetooth keyboard, but there are plenty of other models available too, such as the very portable HiPPiH iEagle Foldable Wireless Keyboard (pictured).

Docks

Apple have launched minimal and functional docks for all their iPhone models, meaning you always have somewhere to park, and charge, your iPhone. The main downside of such stands is that they will not work with any custom cover or case. The current iPhone 4/4S model does have the advantage of having a dedicated line-out, making it easy to connect your iPhone to a stereo for audio playback using an RCA-to-minijack cable.

> **→ TIP** Belkin make very good RCA-to-minijack cables, which happen to come in an iPhone-friendly grey-silver colour scheme. They also produce a matching minijack-to-minijack cable, which can be useful for connecting your iPhone to line-in sockets on computers and speakers.

Airport Express & AirPlay

The more convenient way to play music from your iPhone across a home stereo system is wirelessly, using Apple's AirPlay technology to connect to either an Apple TV (see p.186) or an Airport Express (pictured).

Attach an Airport Express to a power point near your hi-fi and connect it to the stereo with an RCA-to-minijack cable. Then any iPhone connected to the same Wi-Fi network can beam music straight from the Music app to the hi-fi. And what's more, with the Apple Remote app (see p.178) you can use your iPhone as a Wi-Fi remote control for the music coming from iTunes on your Mac or PC.

You can also attach a printer to an Airport Express via USB for wireless printing across your network.

FM radio transmitters

These clever little devices turn your iPhone into an extremely short-range FM radio station. Once you've attached one, any radio within range (theoretically around thirty feet, though a few feet is more realistic to achieve a decent quality of sound) can then tune in to whatever music or podcast the iPhone is playing. The sound quality isn't as good as you'd get by attaching to a stereo via a cable and there can be interference, especially in built-up areas. But FM transmitters are very convenient and allow you to play your iPhone's music through any radio, including those – such as portables and car stereos – that don't offer a line-in socket.

Some models, such as the Griffin iTrip (pictured), connect directly to the Dock connector socket and draw power from the phone's battery. Others contain their own battery and connect via the headphone jack – not as neat, but these types will work with any audio device.

FM transmitters were for a long time technically illegal in Europe, as they breached transmission laws. However, the law has changed, saving iPhone owners the hassle of importing them from the US.

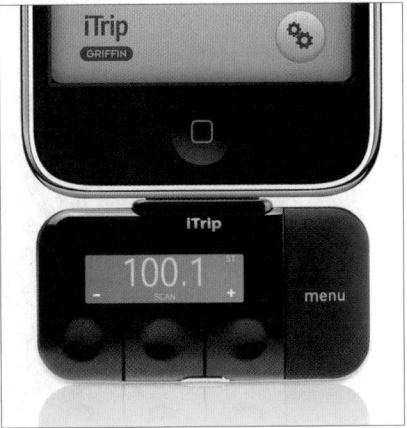

iPhone speaker units

An alternative to connecting your iPhone to a hi-fi or a set of powered speakers is to use a self-contained iPhone/iPod speaker system or clock radio with a built in dock. These are easy to move from room to room (or indeed, from place to place when travelling) and they are space efficient too. There are literally hundreds of models available, many of which can run on batteries for extra portability.

There are audiophile options in all shapes and sizes, from the tiny iHome iHM79, to mid-size systems such as the popular Bose SoundDock. At $300/£200, the latter isn't cheap, but the sound is exceptionally clear and punchy for the size. If you want to get even fancier, take a look at the $600/£670 Bowers & Wilkins Zeppelin Air (pictured).

> ➔ **TIP** For a while, Apple made its own speaker unit, the chunky iPod Hi-Fi, which (given enough batteries to power a small village) could be taken out and about ghettoblaster-style. It was discontinued in 2007, though can still be found on eBay.

If you are after a solid clock radio, that will both wake you up in the morning and charge your iPhone over night, the Phillips DC290 and DC390 systems are both worth a look, and will set you back $100/£150 and $150/£200 respectively.

Car accessories

Several major car manufacturers are now offering built-in iPod/iPhone connectivity – among them BMW, Volvo, Mercedes and Nissan – which will allow you to control your music from your dashboard. Combine this with a Bluetooth headset for safely receiving calls (see p.272) and you're all kitted out. But don't worry if you lack a recent high-end vehicle – you can also do things piecemeal:

Audio connectors

Unless your car stereo has a line-in socket, the two options are an FM transmitter (see p.275) or a cassette adapter. The latter will only work with a cassette player, of course, but they tend to provide better sound than an FM transmitter, and they're not expensive.

Chargers

For in-car charging, try one of the various Incase Car Chargers or the Griffin PowerJolt Charger, both of which slot into a standard 12v car accessory socket.

Cradles

If you want to use your iPhone as a SatNav device, using a TomTom app (see p.232) or similar, you will most definitely need some kind of specialist cradle to mount your iPhone on either the dashboard of windscreen. The Kensington Quick Release Car Mount (pictured) is a good choice at around $30/£20.

Health & fitness

The online Apple Store now has a dedicated section for health and fitness hardware to go with your iOS devices. It's mostly focused on the running market, with the Nike-branded iPod Sensor being the device that has received the most media attention. You pop it into the sole of your running shoe or trainer and it monitors your workouts and running sessions and reports back on your progress via an app.

At the more specialized end of the market are accessories such as the Withings Smart Blood Pressure Monitor (pictured) and WiFi Body Scale, both of which connect with apps to help you monitor your blood pressure and weight respectively.

Accessories online

News and reviews

Everything iCafe everythingicafe.com
iLounge ilounge.com
iPhone Atlas iphoneatlas.com
iPhone Freak iphonefreak.com

Accessory stores

Amazon amazon.com or amazon.co.uk
Apple Store apple.com/store
iLounge ilounge.pricegrabber.com

Index